T0174881

Poor-Quality Cost

QUALITY AND RELIABILITY

A Series Edited by

Edward G. Schilling

Center for Quality and Applied Statistics
Rochester Institute of Technology
Rochester, New York

Additional volumes in preparation

Poor-Quality Cost

H. JAMES HARRINGTON

Chairman, Board of Directors
American Society for Quality Control
Milwaukee, Wisconsin

CRC Press
Taylor & Francis Group
Boca Raton London New York

CRC Press is an imprint of the
Taylor & Francis Group, an **informa** business

First published 1987 by ASQC Quality Press and Marcel Dekker Inc.

Published 2019 by CRC Press
Taylor & Francis Group
6000 Broken Sound Parkway NW, Suite 300
Boca Raton, FL 33487-2742

© 1987 by Taylor & Francis Group, LLC
CRC Press is an imprint of Taylor & Francis Group, an Informa business

First issued in paperback 2019

No claim to original U.S. Government works

ISBN-13: 978-0-367-45151-6 (pbk)
ISBN-13: 978-0-8247-7743-2 (hbk)

This book contains information obtained from authentic and highly regarded sources. Reasonable efforts have been made to publish reliable data and information, but the author and publisher cannot assume responsibility for the validity of all materials or the consequences of their use. The authors and publishers have attempted to trace the copyright holders of all material reproduced in this publication and apologize to copyright holders if permission to publish in this form has not been obtained. If any copyright material has not been acknowledged please write and let us know so we may rectify in any future reprint.

Except as permitted under U.S. Copyright Law, no part of this book may be reprinted, reproduced, transmitted, or utilized in any form by any electronic, mechanical, or other means, now known or hereafter invented, including photocopying, microfilming, and recording, or in any information storage or retrieval system, without written permission from the publishers.

For permission to photocopy or use material electronically from this work, please access www. copyright.com (http://www.copyright.com/) or contact the Copyright Clearance Center, Inc. (CCC), 222 Rosewood Drive, Danvers, MA 01923, 978-750-8400. CCC is a not-for-profit organiza-tion that provides licenses and registration for a variety of users. For organizations that have been granted a photocopy license by the CCC, a separate system of payment has been arranged.

Trademark Notice: Product or corporate names may be trademarks or registered trademarks, and are used only for identification and explanation without intent to infringe.

Visit the Taylor & Francis Web site at
http://www.taylorandfrancis.com

and the CRC Press Web site at
http://www.crcpress.com

Library of Congress Cataloging-in-Publication Data

Harrington, H. J. (H. James)
  Poor-quality cost.

  (Quality and reliability ; 11)
  Bibliography: p.
  Includes index.
  1. Quality control—Costs.   2. Commercial products—Defects—Costs.   I. Title.   II. Series.
  TS156.H346   1987   658.5'62   86-29277
  ISBN 0-8247-7743-3

This book is dedicated to my son, Jim, who has more than lived up to my every expectation, and to my mother, Carrie, who has always encouraged and inspired me to do more and better than I have in the past. Their love and patience have had a major impact on my efforts.

About the Series

The genesis of modern methods of quality and reliability will be found in a simple memo dated May 16, 1924, in which Walter A. Shewhart proposed the control chart for the analysis of inspection data. This led to a broadening of the concept of inspection from emphasis on detection and correction of defective material to control of quality through analysis and prevention of quality problems. Subsequent concern for product performance in the hands of the user stimulated development of the systems and techniques of reliability. Emphasis on the consumer as the ultimate judge of quality serves as the catalyst to bring about the integration of the methodology of quality with that of reliability. Thus, the innovations that came out of the control chart spawned a philosophy of control of quality and reliability that has come to include not only the methodology of the statistical sciences and engineering, but also the use of appropriate management methods together with various motivational procedures in a concerted effort dedicated to quality improvement.

This series is intended to provide a vehicle to foster interaction of the elements of the modern approach to quality, including statistical applications, quality and reliability engineering, management, and motivational aspects. It is a forum in which the subject matter of these various areas can be brought together to allow for effective integration of appropriate techniques. This will promote the true benefit of each, which can be achieved only through their interaction. In this sense, the whole of quality and reliability is greater than the sum of its parts, as each element augments the others.

The contributors to this series have been encouraged to discuss fundamental concepts as well as methodology, technology, and procedures at the leading edge of the discipline. Thus, new concepts are placed in proper prospective in these evolving disciplines. The series is intended for those in manufacturing, engineering, and marketing and management, as well as the consuming public, all of whom have an interest and a stake in the improvement and maintenance of quality and reliability in the products and services that are the lifeblood of the economic system.

The modern approach to quality and reliability concerns excellence: excellence when the product is designed, excellence when the product is made, excellence as the product is used, and excellence throughout its lifetime. But excellence does not result without effort, and products and services of superior quality and reliability require an appropriate combination of statistical, engineering, management, and motivational effort. This effort can be directed for maximum benefit only in light of timely knowledge of approaches and methods that have been developed and are available in these areas of expertise. Within the volumes of this series, the reader will find the means to create, control, correct, and improve quality and reliability in ways that are cost-effective, enhance productivity, and create a motivational atmosphere that is harmonious and constructive. It is dedicated to that end and to the readers whose study of quality and reliability will lead to greater understanding of their products, their processes, their workplaces, and themselves.

Edward G. Schilling

Foreword

When the fact that good quality goes hand-in-hand with good cost becomes engrained in the conviction and action of every man and woman in a company, it provides the organization with a basic foundation for its competitive leadership. It brings an essential company-wide strength for successfully serving today's markets, in which the way to make products and services quicker and cheaper is to make them better.

One of the most damaging managerial myths of the past was the belief that better quality required higher cost, and would somehow make production more difficult. Nothing could have been farther from the facts of business experience. These facts have repeatedly demonstrated the basic principle that good quality means good resource utilization—of equipment, of materials, of information, and above all of human resources—and consequently means lower costs and higher productivity.

Our original development of the concept and quantification of quality costs has had the objective of equipping men and women throughout a company with the necessary practical tools and detailed economic know-how for identifying and managing their own costs of quality. These costs have successfully provided the common denominator in business terms both for managing quality as well as for communication among all who are involved in the quality process. Therefore, we have continued to develop, implement, and refine the cost of quality in companies the world over.

Dr. H. James Harrington, in *Poor-Quality Cost,* adds significant value to the development of the economics of quality and its utilization and management. He brings to this work many years of successful quality experience, which is reflected both in the basically practical approach of the book and in the clarity of its examples and illustrations. Jim Harrington's inherent professionalism and his long-demonstrated commitment to serve others not only in the United States but throughout the world provide *Poor-Quality Cost* with a conviction, a credibility, and a positive spirit that help the book spring to life for the application benefit of the reader.

This book is a most welcome contribution to the literature and the know-how of the modern quality field. It is and will be both useful and productive for men and women interested in quality the world over.

Armand V. Feigenbaum
President and
Chief Executive Officer
General Systems Co., Inc.
Pittsfield, Massachusetts

Preface

For years, management believed that it was more expensive to provide high-quality products and service to customers, and used this excuse to keep the company's output from reaching its full potential. During the 1970s and 1980s, management's attitude began to change as they found that in international markets quality products provided greater return on investment and increased the company's market share. As a result, a great deal of attention was focused on improving the quality of output from all employees. This increased focus revealed three truths:

1. It is not more expensive to provide high-quality products and services. In fact, in many cases, it is less expensive.
2. When the quality problems are solved, cost and schedule problems are greatly reduced. Management

must therefore put quality first in every decision they make. As the old saying goes, "The bitterness of poor quality lingers long after the sweetness of meeting schedule."

3.   The terms used by most quality professionals are completely foreign to management and are difficult, if not impossible, to summarize in a company-wide unit of measure that can be used effectively by management. As a result, terms such as *percent defective, throughput yields, defects per unit,* and *mean time to failure* have been translated into a common denominator—dollars—to manage the business.

To solve the third problem, Armand V. Feigenbaum, while working at General Electric Company in the early 1950s, developed a dollar-base reporting system called "quality cost." This system pulled together all the costs related to developing the quality system and inspecting products, as well as the cost incurred when the product failed to meet requirements. He then provided management with a report that got their attention—one that was based on dollars, the language of top management and the stockholder. Over the years, Dr. Feigenbaum's quality-cost concept has been refined and expanded to the point that today it provides an excellent management tool that can be used to direct quality-improvement activities and measure the effectiveness of the total quality system.

Since their inception, quality-cost systems of one kind or another have been implemented by many companies to help management direct their improvement activities and measure the quality system's effectiveness. Among these companies are: IBM, Bendix, Komatsu Ltd., Abbott Laboratories, Westinghouse, Honeywell, General Electric, International Telephone & Telegraph, Irving Trust, Eaton, Digital Equipment, Allis-Chalmers Canada, General Motors, and many more.

Unfortunately, the term "quality cost" leaves a negative impression that reflects the thinking of the 1950s, when it was believed that better-quality products cost more to produce. Given the change in management attitude toward quality, and the new dimensions that have been added to the original concept, the term "poor-quality cost" (PQC) seems more appropriate and will be used for the concept that is presented in this book. As you will see, the system—whether it is called quality cost or poor-quality cost—is designed to help reduce the cost associated with poor quality.

Poor-quality cost varies among companies; these variations are based on product complexity, state of the technology used, how the customer uses the product, the elements of PQC that are included, and the level of refinement of the quality system within the company. In many cases, PQC accounts for more than 40% of sales price. IBM, for example, has reported that their PQC ranged between 20 and 40% of revenue before they started their quality-improvement process. This is not out of line when a high-technology company considers both the blue- and white-collar PQC.

Most company presidents accept that poor quality is costing them a great deal of money, but they are absolutely shocked when they find out what the cost really is. John Akers, President of IBM, reported, "When we analyzed what we were spending on quality—the cost to make things right, as well as the cost to fix and rework things that weren't right—we were surprised and disturbed. We found that our total quality costs were higher than we had thought. Roughly one-quarter was what we call prevention and appraisal cost, and roughly three-quarters were failure costs."

Other companies have found costs in the same range. James E. Preston, President of Avon, noted when discussing quality costs, "The cost of building quality into the product is 5 percent of sales, while the cost of nonconformance is 20 percent."

The poor-quality-cost concepts presented in this book differ from the original quality-cost concepts in the following ways:

1. To help win acceptance of the concept in the white-collar areas, the term "defects" has been replaced with "errors."
2. In keeping with the current needs for continuous improvement, the concept of an optimum quality-cost operating point has been changed to reflect the cost advantages of error-free performance.

3. Test equipment costs are taken out
    of the appraisal category and placed
    in a separate category so that these
    costs can be spread equally across
    the total output they support.
4. The cost that the customer incurs as
    a result of poor quality is taken into
    consideration.

The purpose of this book is to help your
company develop a poor-quality-cost system that will provide management and the
employers with data that can be used to
identify improvement opportunities, optimize the effectiveness of the improvement
efforts, and measure the progress that is
being made by the improvement process.

H. James Harrington

Contents

1

Why Poor-Quality Cost?

What Is Poor-Quality Cost? Poor quality costs your company money. Good quality saves your company money. It's as simple as that. James E. Olson, President of AT&T, said, "A lot of people say quality costs you too much. It does not. It will cost you less" (1). But many companies today do not measure the cost of poor quality, and if you do not measure it, you cannot control it. Why is it, then, that those in corporate management do not insist on the same good financial control over poor-quality cost (PQC) that they exercise over the purchase of materials, when often PQC exceeds the total materials budget?

This book discusses the cost of not having quality—not quality cost. It is often cheaper to provide high-quality products and services than to provide shabby ones. Quality is not the cost of providing an output. It is the value the customer receives from the output. Ronald Reagan wrote, "Consumers,

by seeking quality and value, set the standards of acceptability for products and services by 'voting' with their marketplace dollars, rewarding efficient producers of better
quality products and performance" (2).
Donald E. Peterson, Chairman of the Board
of Ford Motor Company, stated, "World-
class quality means providing products and
services that meet customer needs and expectations at a cost that represents value to
the customer" (3). Of course, it is not necessary to produce products or services that
greatly exceed the customer's expectations,
but it is always necessary to fully meet
those expectations. It is almost as wasteful
to produce paper cups that leak as it is to
produce silver-plated drinking cups that
will be thrown away after one use. We need
to have a system that will define the difference between luxury and fitness for use,
between waste and optimum performance.
Part of this system is a quantification of
what your company is spending because all
things, people, and materials are not per-
fect. To put it simply, the poor-quality cost-
reporting system is only one of the many
tools needed in a comprehensive, company-
wide quality system, but it is an important
tool in that it directs management attention and measures the success of the company's efforts to improve. It also provides
management with the necessary tools to
ensure that suboptimization does not have
a negative effect on the total system. The
importance of poor-quality cost was recog-
nized by the United States Department of
Defense when a requirement for PQC sys-

tems was included in Military Standard MIL-Q-9858A.

In the country of Utopia, poor-quality costs are zero. The workers always assemble parts correctly so there is no need to test anything. There is never a flaw in the materials, and the products always work perfectly.

But here in the United States, and in most other places in the world, things are a bit different. People make errors, equipment malfunctions, parts break down—even Ivory soap is only 99.44 percent pure. As a result, we need testers, inspectors, repair people, and complaint departments. The difference between the two countries in producing, maintaining, and owning a product illustrates the definition of poor-quality cost.

Poor-quality cost is defined as all the cost incurred to help the employee do the job right every time and the cost of determining if the output is acceptable, plus any cost incurred by the company and the customer because the output did not meet specifications and/or customer expectations. Table 1.1 lists the elements of poor-quality costs. Each will be discussed in detail later in this book.

Where Is PQC Used?

During the 1960s and '70s, poor-quality cost was used primarily to measure manufacturing and warranty costs, but in re-

TABLE 1.1   The Elements of Poor-Quality Costs

 I. Direct poor-quality costs

    A.   Controllable poor-quality cost

        1.   Prevention cost

        2.   Appraisal cost

    B.   Resultant poor-quality cost

        1.   Internal error cost

        2.   External error cost

    C.   Equipment poor-quality cost

 II. Indirect poor-quality costs

    A.   Customer-incurred cost

    B.   Customer-dissatisfaction cost

    C.   Loss-of-reputation cost

---

cent years management has realized that all departments (both blue-collar and white-collar) make errors. Numerous studies have been made that show that white-collar poor-quality cost accounts for 20 to 35 percent of the total effort expended by these departments. In most companies, the cost of administrative errors and the resulting checks and balances are accepted as a way of life. Applying poor-quality cost to the white-collar areas focuses management attention on this neglected waste.

It is also necessary to apply poor-quality cost systems to the impact that errors have on the customer. Frequently, the cost incurred by the customer when an error occurs can far exceed the cost of repairing the defective item. Consider a 10-year-old boy who is delighted to find a new red and white bicycle beneath the Christmas tree. When he and his father try to assemble the bicycle, everything goes well until they attempt to put on the front wheel and find that a nut is missing. As a result, before the bicycle can be used, the father must make a trip to the bicycle store, wait in line to get a new nut, and return home—a waste of one hour of valuable time and 24 miles of travel. The cost to the company is a 5¢ nut; the cost to the customer is 300 times more.

Why Use PQC?

Poor-quality cost provides a very useful tool to change the way management and employees think about errors. PQC helps by:

1. Getting management attention—talking to management in dollars provides them with information that they relate to. It takes quality out of the abstract and makes it a reality that can effectively compete with cost and schedule.

2. Changing the way the employee thinks about errors—when as a result of an employee's actions a defective gear is scrapped, there will be greater impact on his or her future performance if the employee knows it costs $100. In one case, what is thrown away is only a piece of metal;

in the other case, it's a $100 bill. Employees need to understand the cost of errors they make.

3. Providing better return on the problem-solving efforts—poor-quality cost "dollarizes" problems so that corrective action can be directed at the solutions that will bring maximum return. James R. Houghton, Chairman of Corning Glass Works, has reported, "At Corning, cost of quality is being used to identify opportunities, to help prioritize those opportunities, and to set targets and measure progress. It's a tremendous tool, but we are taking great care to ensure that it is not used as a club" (3).

4. Providing a means to measure the true impact of corrective action and changes made to improve the process—by focusing on poor-quality cost of the total process, suboptimization can be eliminated.

5. Providing a simple, understandable method of measuring what effect poor quality has on the company and providing an effective way to measure the impact of the quality-improvement process.

**PQC
Limitations**

Poor-quality cost by itself cannot resolve your quality problems or optimize your quality system. It is only a tool that helps management understand the magnitude of the quality problem, pinpoints opportunities for improvement, and measures the progress being made by the improvement activities. The PQC system must be accompanied by an effective improvement process that will reduce the errors that are being made in both the white- and blue-collar areas. For information on how to implement an effective improvement process, read *The Improvement Process—How America's Leading Companies Improve Quality and Productivity,* by H. James Harrington, jointly published by McGraw-Hill and the American Society for Quality Control. (See "Suggested Additional Reading," page 189.)

2

Understanding Direct Poor-Quality Cost

Of the two major poor-quality cost categories, direct and indirect, the direct PQC are the better understood and are traditionally used by management to run the business because the results are less subjective. Direct poor-quality costs can be found in the company ledger and can be verified by the company's accountants. They include all the costs a company incurs because management is afraid that people will make errors, all the costs incurred because people do make errors, and the costs related to training people so they can do their jobs effectively. Direct poor-quality costs encompass three major types of expenditure: controllable PQC, resultant PQC, and equipment PQC (see Table 2.1). Each of these will be discussed in the following pages.

TABLE 2.1   Direct Poor-Quality Costs

A.   Controllable poor-quality cost
    1.   Prevention cost (investment in prevention)
    2.   Appraisal cost

B.   Resultant poor-quality cost (loss)
    1.   Internal error cost (loss)
    2.   External error cost (loss)

C.   Equipment poor-quality cost

Controllable PQC

Controllable poor-quality costs are those that management has direct control over to ensure that only customer-acceptable products and services are delivered to the customer. Controllable poor-quality costs are further subdivided into two categories: prevention costs and appraisal costs.

Prevention Costs

Prevention costs are all the costs expended to prevent errors from being made or, to say it another way, all the costs involved in helping the employee do the job right every time. If you look at this from a financial viewpoint, it is really not a cost. It is an investment in the future, often called a cost-avoidance investment.

Typical prevention costs are those of:

- developing and implementing a quality data-collecting and reporting system

- developing the quality process control plan
- quality-related training
- job-related training
- making vendor surveys
- implementing the improvement process
- conducting design concept reviews
- preventing a problem from recurring (preventive action)

(For more examples see the Appendix.)

By far the best way a company can spend its poor-quality cost dollars is to invest in preventive action. Unfortunately, however, most companies have neglected this valuable investment because it is difficult to tie it to a tangible return on the investment. IBM President John F. Akers, when talking about IBM's plan to improve quality during the 1980s, said that during the previous decade, "We were responding to problems as they surfaced instead of doing everything possible to prevent them from occurring in the first place" (4). Today, many companies are moving from a reactive business system to one based on prevention.

Appraisal Costs   Appraisal costs are the result of evaluating already completed output and auditing the process to measure conformance to established criteria and procedures. To say it another way, appraisal costs are all the costs expended to determine if an activity was done right every time. Typical appraisal costs are those of:

- quality-assurance audits of the manufacturing process

- outside financial audits

- inspection and testing to determine conformance of products and/or services to specifications

- inspection of purchased material, both in plant and at the supplier's location

- approval signatures on a document

- outside endorsements, such as from Underwriters Laboratories

- maintenance and calibration of test and inspection equipment

- review of completed designs

- review of test and inspection data

- second-level managers' review of first-level management decision

- proofreading letters

- quality data processing and reporting

- payroll audits

- field performance testing

- certification evaluation (example: lawyers taking the bar exam)

(For more examples see the Appendix.)

The only reason you need appraisal cost is that often management is not confident that the money and time expended in prevention cost are 100 percent effective at eliminating the possibility of error. Often appraisal activities are too late and too little.

Impact of Changing Prevention and Appraisal Costs

To simplify the definition, we can say that preventive activities are those activities that have a positive effect on a person's ability to do the job right every time or, in other words, activities that improve first-time yield. As we increase the preventive activities, we reduce the total error cost because the total number of errors is reduced (see Figure 2.1). This occurs because management has provided the employees with the training, tools, equipment, systems, and knowledge to enable them to do the job right a high percentage of the time.

Appraisal activities, on the other hand, prevent errors from being delivered to the customer or to a higher level of assembly. Appraisal activities do not reduce the total number of errors; they only detect a higher percentage of errors in the output before it is delivered to the company's customer. Figure 2.2 shows that the total error costs and the total number of errors remain constant, even though a great deal of additional money is invested in increased appraisal.

For simplification, the graph shown in Figure 2.2 assumes that the cost of repairing an error in the manufacturing operation is the same as it is after the output has been delivered to the customer, but in most cases

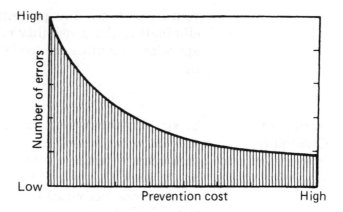

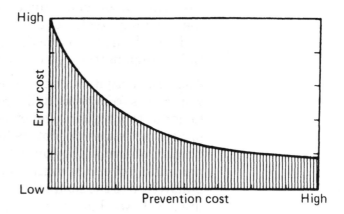

FIGURE 2.1 Effect of prevention cost on the
total number of errors and the total error cost.

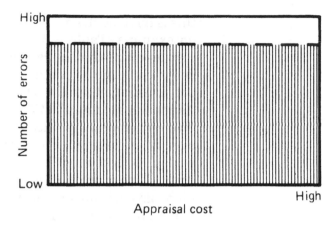

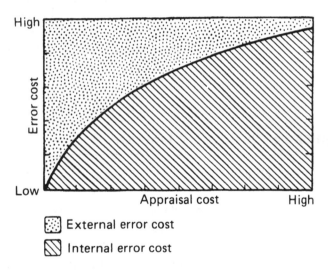

FIGURE 2.2 Effect of appraisal cost on the total number of errors and total error cost when the individual external and internal error costs are equal.

this is not true. Using a typical computer as an example, Table 2.2 shows the quality-cost leverage that is realized by finding a defect early in the process, before the product is shipped to the customer. John F. Akers, President of IBM, put it this way: "We have found significant financial leverage by investing in prevention and appraisal, both of which greatly reduce failure costs. Some of our divisions show extremely high payback on prevention investments, in both hardware and software. Fixing it in the lab before it reaches the field is where the payoff is" (4). Studies at Hewlett-Packard revealed that a defective resistor costs 2¢ if thrown away before use. It costs $10 if found at the board-assembly level, and hundreds of dollars if it is not discovered until it reaches the customer.

TABLE 2.2 Quality-Cost Leverage

Hardware		Software	
When corrected	Relative cost impact	When corrected	Relative cost impact
Component design	Negligible	Design/code	1×
Subassembly	1×	Internal test	20×
Unit	10×	After delivery to customer	80×
Field	50×		

Figure 2.3 shows the effect on total error costs if the cost of the external error is two times the cost of the internal error. You will note that as appraisal costs increase, total error costs decrease rapidly, even if the total number of errors remains constant.

It is easy to see that the only reason we need appraisal activity is that the prevention activity may not be wholly effective. After the appraisal system has defined the problem, it is imperative that corrective action be immediately implemented to prevent it from recurring. How many times have you worked on and solved a problem that shut down the line or impeded office efficiency, only to have it come back and bite you again, perhaps six months, a year, or even three years later? One IBM manager was heard saying to a group of inspectors, "We have to look at each error we detect and be sure that that problem is put to bed." He was interrupted by an inspector who pointed out, "If you put something to bed, it can get up and bite you again. What you really need to do is to bury the problem so that it will never come back." She was right.

All too often, we correct the problem that has stopped the process but do not take the time to analyze why it occurred and take action to assure that the process has been permanently changed so the problem will not recur. It is only when we have implemented action that will prevent the problem from recurring that the problem is really solved.

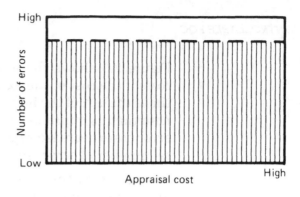

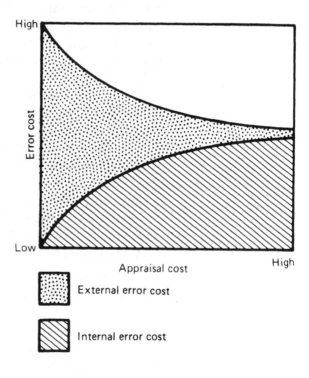

FIGURE 2.3 Effect of appraisal cost on total number of errors and total error cost when the individual external error cost is two times internal error cost.

22

Resultant PQC

Resultant poor-quality costs make up the second category of direct PQC. They include all the company-incurred costs that result from errors or, to put it another way, all the money the company spends because all activities were not done right every time. These costs are called resultant costs because they are directly related to management decisions made in the controllable poor-quality cost category. Resultant costs are divided into two subcategories: internal and external error costs. The items in this category could appropriately be called losses rather than costs, for they truly are direct losses to the company.

Internal Error Cost

Internal error cost is defined as the cost incurred by the company as a result of errors detected before the output is accepted by the company's customer. In other words, it is the cost the company incurs before a product or service is accepted by the customer because everyone did not do the job right every time. Included are the costs incurred from the time an item is shipped from a supplier until it has been accepted by the final customer. The following are examples of internal error costs:

- in-process scrap and rework
- retyping letters
- troubleshooting and repairing
- engineering changes
- material review board activities

- additional costs because bills were paid late

- costs resulting when additional inventory is required to support poor process yields, potentially scrap parts, and rejected lots

- reinspection and testing after an item has been found defective

- downgrading (some lower-quality products are classified as seconds and sold at reduced cost)

- computer reruns

- overcooked food

(For more examples see the Appendix.)

External Error Cost

External error cost is incurred by the producer because the external customer is supplied with an unacceptable product or service. It is the cost incurred by the company because the appraisal system did not detect all the errors before the product or service was delivered to the customer.

Typical external error costs are related to items such as the following:

- customer-rejected services or products

- product liability suits

- complaint handling

- warranty administration

- repair-personnel training

- stocking of parts to support field repair

- handling of returned material and repair of defective products that have been delivered to the customer

- product recall or updates in the field

- overhead required to maintain field service centers

(For more examples see the Appendix.)

Equipment PQC

The last kind of direct poor-quality cost is equipment PQC. Investment in equipment used to measure, accept, or control the product or service, plus the cost of the space that equipment occupies, make up equipment PQC. This includes the cost of the equipment used to print and report quality data. Examples are computers, typewriters, voltmeters, micrometers, coordinate measuring machines, and automated test equipment. In some cases, environmental controls installed to reduce the possibility of making errors are included in equipment poor-quality cost (for example, sound baffles, clean rooms, and air-conditioning controls). Equipment poor-quality cost does not include equipment used to make the products, such as lathes, drill presses, and assembly fixtures; electrical equipment required to adjust the product so it will perform to specifications; or computing systems used for accounting and/or scheduling.

3

Direct Poor-Quality-Cost Curves

Interaction Between Controllable and Resultant PQC

To better understand poor-quality cost, let us study the theoretical interaction between controllable and resultant PQC (Figure 3.1). On the left side of the curve, controllable poor-quality cost is very low. This causes the resultant PQC to be very high, because there is little money being spent to prevent errors or to detect them before they are delivered to the customer. As the controllable PQC increases, resultant costs are decreased because fewer errors are made and more errors are detected before the output is delivered to the customer. At the right side of the curve, although controllable poor-quality cost is significantly increased, there is a negligible decrease in resulting costs because increasing controllable PQC becomes less and less effective. Consider this: The first time you proofread a report, you may find four errors. The second time, you may find only one. Then, if you proofread it another five times, you may find only one additional error.

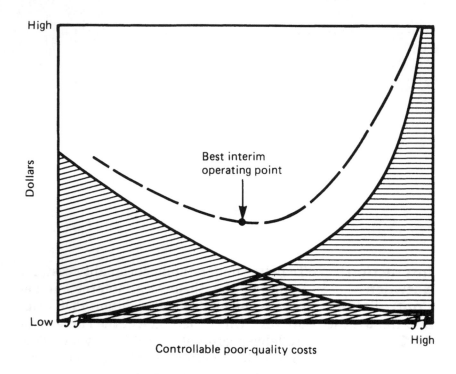

Controllable poor-quality costs

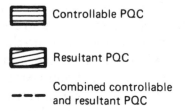

FIGURE 3.1 Effect of varying controllable poor-quality costs.

**Best Interim
Operating
Point**

When the controllable and resultant costs are added together, a new curve is developed. This curve shows a picture of the costs that result from the interaction between controllable and resultant poor-quality costs. An effective quality system should operate at the point on the curve labeled "best interim operating point." At this point, the total controllable and resultant PQC are minimized and the return on investment is maximized for that point in time. The term "best interim operating point" was chosen very carefully: It is the best point for one set of conditions only and should continue to change as the improvement process drives the error level lower. You will also see later that this operating point will change as we consider indirect poor-quality cost.

Although this curve shows the interaction between controllable and resultant PQC, its actual values are correct only for a single point in time. Later, in Chapter 4, you will see how controllable and resultant poor-quality costs work to complement each other, causing a continuous decrease in both expenditures and proving the old saying, "It's always cheaper to do it right the first time than to do it over."

It is obvious by now that the quality-cost picture is a complex one with many interacting elements. For example, an increase in preventive expenditures can reduce appraisal costs and internal and external error costs because there are fewer errors in the output. Adding additional automated test equipment can increase equipment poor-quality cost and internal error cost, and at the same time decrease appraisal cost and external error cost.

To make these complex interactions understandable, it is useful to portray poor-qual-

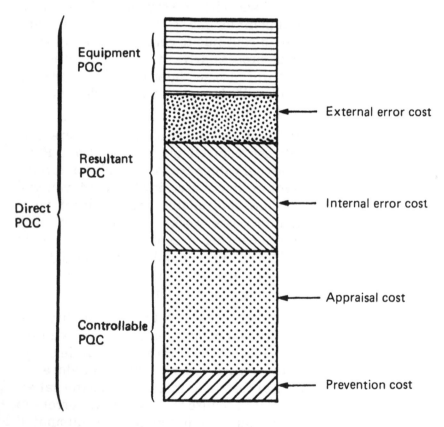

FIGURE 3.2 Direct poor-quality cost.

ity cost graphically. Direct PQC for any one time period can be illustrated with a simple bar graph. Equipment PQC has been added to Figure 3.2. The bar graph thus portrays the total direct PQC. Normally, equipment costs are spread equally across all units manufactured for the life of the program, so at times these costs are dropped off the individual product graph. This is poor practice, however, as it makes product-to-product comparisons invalid even if all other factors are the same.

A word of caution about comparing two or more product lines: Just as products are different, poor-quality costs can be different. A manufacturer of simple hardware can and should have much lower PQC than a company that produces more complex electronic components. It is obvious that companies manufacturing a sophisticated product where stringent reliability specifications require more complex quality systems will have higher poor-quality costs. In these cases, prevention, appraisal, and internal error costs tend to be on the high side, while external error costs are extremely low.

Percentage of Value Added   PQC for different time periods or different products can be compared by plotting PQC per item divided by total sales value of the products produced during the time period (poor-quality cost as a percent of sales price). Although percentage of sales price is the most common method used to report PQC, an even better way is percentage of value added. Percentage of value added provides a weighting factor for the differences in complexity from product to product, and allows for the fact that poor-quality costs incurred by suppliers are not included in the PQC data available to most companies. Obviously, poor-quality cost as a percentage of sales will be much lower for a product that has 95 percent of the sales cost vended, compared to a product that is totally manufactured within the company.

Changes in PQC with Time

A series of bar graphs reflecting different time periods provides a means of modifying controllable factors and measuring their impact upon the total PQC system.

The difference between time periods A and B in Figure 3.3 is that B has more preventive expenditure and less appraisal cost. As a result, the overall direct poor-quality cost decreased during the second time period. In time period C, appraisal cost was increased compared to time period B, which resulted in a reduction in external error cost greater than the increase in internal error cost, causing the total PQC to decrease. In time period D, appraisal cost was reduced to its lowest level. This reduced internal error cost but greatly increased external error cost, resulting in an increase in overall direct PQC. Resultant PQC normally will go down as controllable poor-quality costs are

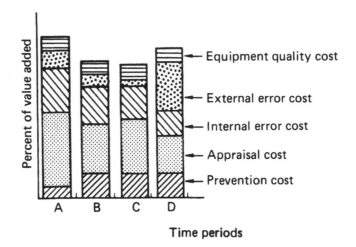

FIGURE 3.3 Effects of modifying controllable poor-quality cost on same product.

increased. The decrease in resultant poor-quality cost may be greater or less than the change in controllable PQC, depending on the previous quality system's level of sophistication.

Richard K. Dobbins, Staff Quality Engineer at Honeywell, Inc., reports, "It has not been uncommon for companies to reduce their failure (error) losses by about $9.00 for every $1.00 invested as a quality appraisal expense, and reductions of up to $15.00 in failure losses have been experienced for every $1.00 invested in quality prevention expenses" (5).

4

Direct Poor-Quality-Cost Analysis

Why Spend Dollars on Prevention?

Before we can embark on a program to reduce poor-quality cost, we need to understand how its elements interact with one another. This is the place where a PQC analysis system can most effectively be used. Figures 4.1 and 4.2 portray two standard poor-quality cost curves for the same product. The internal and external error cost levels for both graphs are the same at zero controllable PQC, but from that point on the curves take drastically different shapes. Note how internal error costs increase as appraisal costs increase on the left side of the graphs, but the increase is more than offset by the decrease in external error costs. Controllable poor-quality cost is the same in both cases, but in Figure 4.1 (Case 1), the controllable PQC is applied largely to appraisal costs, and in Figure 4.2 (Case 2), a large portion of the controllable poor-quality costs are applied to prevention activities.

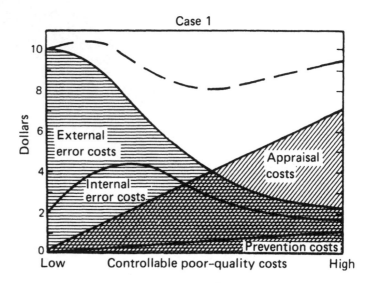

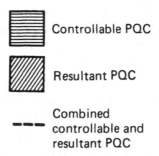

FIGURE 4.1 Case 1 with major expenditures in appraisal activities.

Prevention costs are expenditures designed to help employees do the job right every time. It is therefore obvious that spending quality dollars in doing the job right every time decreases internal and external error

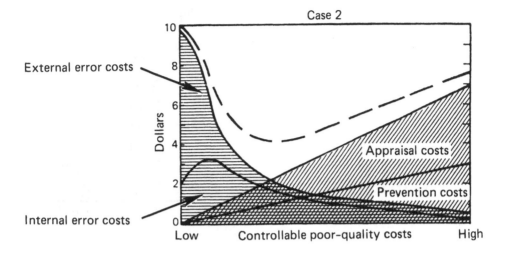

FIGURE 4.2 Case 2 with increased expenditures in prevention activities.

rates. At the same time, appraisal costs can be decreased because:

- Inspection levels can be reduced because the quality of the work entering the inspection station has improved.

- Less inspection time is required to reinspect rejected lots because fewer lots are rejected.

Figure 4.3 shows two bar graphs that represent the product poor-quality cost for Case 1 (Figure 4.1) and Case 2 (Figure 4.2) at their best interim operating points. These bar graphs clearly show the potential savings available from a prevention-based system as opposed to an appraisal-based sys-

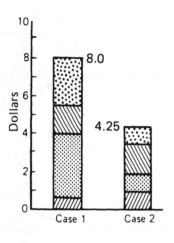

FIGURE 4.3 Best interim
quality cost operating point.

tem. In this case a typical process PQC was
cut from $8 million to $4.25 million. O. G.
Kolacek, Manager of Quality Assurance for
Allis-Chalmers Canada, Limited, reported
that over an eight-year period the company
directed more of this cost to prevention and
appraisal activities in order to reduce total
quality costs, with the following results (6):

	Prevention	Appraisal	Internal errors	External errors
Year 1	5%	13%	36%	46%
Year 8	26%	35%	33%	6%

Not only was the direct poor-quality cost reproportioned, but the total poor-quality costs were reduced by 70 percent (6).

Interaction Between Prevention and Appraisal Activities

At the start of any program, the only poor-quality-cost element is prevention; before long, designs are complete and production begins. With this increased activity, appraisal costs jump. Early in the product cycle, both appraisal and prevention costs run very high—and rightly so. It is the time in the product cycle when problems are the least expensive to correct and have the biggest payback to the program. Problems found in the high-production stage of the product cycle bring major scrap and rework bills, not to mention customer dissatisfaction and loss of reputation. For these reasons, a very aggressive appraisal program should be implemented early in the product cycle.

At the same time, a comprehensive program of error prevention and corrective action must parallel the appraisal activities to make sure that maximum use is made of the data collected during the appraisal operations. This combination puts into motion a complementary cost-reduction system. Errors are systematically tracked down and action is taken to prevent them from recurring. This causes the error level to decrease, allowing 100 percent screening operations to be replaced with sampling. As the appraisal operations continue, there are fewer problems that need to be tracked down to their source, allowing cutbacks in the corrective-action and error-prevention activities. As more and more problems are eliminated, the sampling plans reject fewer and fewer parts, allowing the sample inspection operations to be replaced with audits, and on and on it goes. As a result, the controllable PQC factor is reduced in a step-by-step fashion (see Figure 4.4).

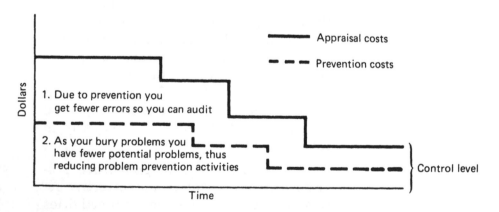

FIGURE 4.4 Prevention/appraisal interaction.

Eugene J. Eckel, Vice President of AT&T, while discussing poor-quality cost, said (7):

> Our experience is that the first cost to come down with the introduction of quality improvement is failure costs. Then, as product quality continues to improve, you'll see inspection costs start to shrink as quality gains continue. At the same time, the investment in prevention will increase, although the overall costs will be reduced. Our cost system is how we arrived at the $30 million-plus savings for the first five months of 1985.

PQC Versus Time

By properly using controllable poor-quality costs and improvement-process resources, you can substantially reduce the resultant PQC (improved product quality), allowing you to reduce controllable PQC. The result is that you have improved product quality and at the same time reduced poor-quality cost. A good quality program is obviously more profitable to a company than a shabby one, but it requires an up-front investment by management (see Figure 4.5). The payback to the corporation in the long run is reliable, high-quality products supported by a less expensive poor-quality cost system.

PQC Versus Product Cycle

Figure 4.6 represents a poor-quality cost life cycle for a typical product. It starts in development engineering when no hard-

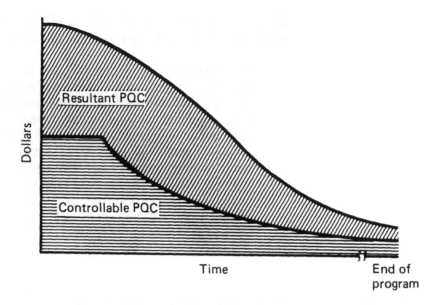

FIGURE 4.5   Poor-quality costs versus time.

ware is being produced. At this point in the cycle, the total PQC effort is expended on the prevention element. As the development pilot line begins to produce products to be subjected to the development test evaluation cycle, we start to see appraisal and error costs. All the poor-quality-cost elements remain high as we enter into the product stage, where equipment is debugged and certified, operators are trained, and operating procedures are debugged and verified. It is a time of low productivity and high poor-quality cost, a period of learning.

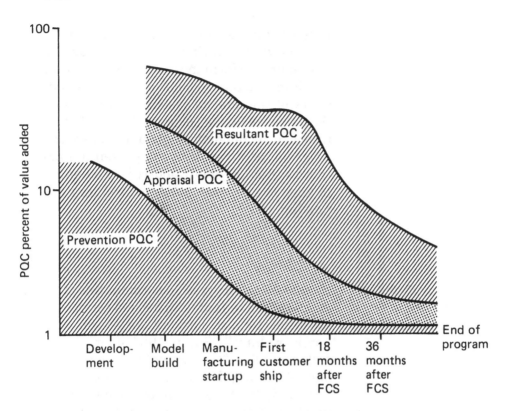

FIGURE 4.6 Product cycle impact on poor-quality cost.

As we enter the production stage and start to manufacture products for manufacturing tests, the efforts of the improvement process begin to take hold, helping to reduce resultant poor-quality cost. As we start shipping to our customer, we see that improvement activities are having a major impact on reducing resultant PQC and appraisal costs, thereby increasing manufacturing throughput capability. During the

next 18 months, the appraisal and prevention activities continue to decrease as the
customer performance data is received,
providing increased confidence in the adequacy of the quality system. You will note
that the resultant poor-quality cost continues to decrease throughout the life of the
program, as a result of the prevention activities that are continuously implemented
to keep the process under control and improving.

**Poor Quality
Is Poor
Business**

Ronald Reagan wrote, "A commitment to
excellence in manufacturing and services is
essential to our nation's long-term economic welfare" (8). In 1984 John Akers
called quality "the competitive edge" (4).
John A. Young, President of Hewlett-Packard, said, "In today's competitive environment, ignoring the quality issue is tanta-
mount to corporate suicide" (9). When it
comes to return on investment, the best
opportunity most companies have today
lies in expanding their efforts to improve
the quality of their products and/or services, because there is a fourfold payback
from improved quality:

1.   Parts that were previously scrapped
     are now usable. For example, con-
     sider an integrated-circuit manufac-
     turer who is experiencing a 40
     percent yield from a state-of-the-art
     component. If the manufacturer
     could improve the throughput yield

to 50 percent, it would realize a 25 percent increase in manufacturing capability and reduced manufacturing costs per unit.

2. People and equipment used primarily in doing rework are now free to make additional good shippable products, thus further increasing the production capability of the manufacturing process without increasing the labor force or the capital equipment. Suppose 20 percent of your manufacturing labor hours have been spent doing rework and, through an effective improvement process, you were able to reduce these hours to 10 percent. The labor, equipment, and space that were previously devoted to doing rework could now be applied to manufacturing additional products, thus increasing the production line's capability by an additional 10 percent plus 1 percent without additional expenditure. The increase is 11 percent, not 10 percent, because there is an additional 1 percent increase due to the reduced inherent error rate in the process.

3. As product quality improves, it becomes less necessary to expend large amounts of appraisal dollars to ensure that the customer's requirements are being met. This can affect the process in many ways. It can result in the use of:

   • Sampling plans in place of 100 percent screening

- Skip lot inspection in place of lot-by-lot acceptance

- Periodic auditing of the product/process in place of sampling

- Process control monitoring in place of product inspection

- Reduced time required during the performance run evaluations to gain the same level of confidence

Whatever the impact on the process, the results are all the same—a reduction in appraisal effort accompanied by large dollar savings.

4. Quality becomes a marketing and sales weapon. If American industry has learned nothing else from the Japanese, it has learned that today's sophisticated customers look at more than just the purchase price when they make their decisions to buy. As your quality reputation increases, so do your share of the market and your profits. On an average, the return on investment for high-quality products is at least 25 percent greater than for low-quality products.

Figure 4.7 shows an operating curve for a typical electronic-equipment manufacturing process. To evaluate the best interim operating point, let us assume a production line is operating at Point A on the combined controllable and resultant poor-quality-cost curve. Operating at Point A, the assembly and test lines are able to produce 2000 units per week at an average profit of $6 per

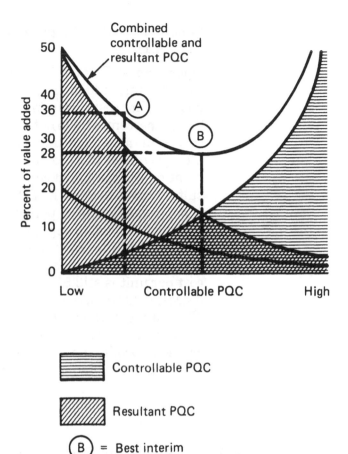

FIGURE 4.7 Comparison of two operating points. B represents the best interim operating point.

unit, or a weekly total profit of $12,000 (see Table 4.1).

Operating at Point A, the basic manufacturing value added is $16 per unit, and poor-quality cost accounts for an additional $9 (36 percent of $25) of the value-added cost. If management increases the controllable factors of PQC to bring the operating point on the combined controllable and resultant curve to Point B (Table 4.1), there is a net reduction of 8 percent in the total value-added cost, dropping the value-added cost from $25 to $23 (92 percent × $25 = $23). The basic manufacturing value-added cost per unit is still $16, but the poor-quality cost value added per unit has dropped to $7. As a result the profit per unit increased from $6 to $8 per unit. This 33 percent increase in profit was achieved simply by increasing controllable poor-quality cost to

TABLE 4.1 Analysis of Two Operating Points on the Poor-Quality Cost Curve

	Point A	Point B
Material cost per unit	$11	$11
Value-added cost per unit	$25	$23
Sales price	$42	$42
Profit per unit	$6	$8
Profit improvement of Point B over Point A		33%

optimize the combined controllable and re-sultant PQC at one point in time, but this is only a small part of the increased profit that can be potentially realized from a PQC system and a quality-improvement process. The real savings result from the interaction of reducing error levels and the accompanying reduction in appraisal levels.

In Figure 4.8 the best interim operating point moved to the right and down as error levels improved, allowing appraisal cost to be reduced, reaching Point C. Where additional reductions cannot be justified, poor-quality cost will never drop to zero because some prevention activities (for example, training) should always be part of the program and some appraisal costs are needed to provide management with the assurance that the process or service can meet, and is meeting, customer expectations.

Reducing poor-quality cost from Point A to Point C (Figure 4.8) increases profits by more than 250 percent (see Table 4.2). At operating points A and C, the basic manufacturing value-added costs are $16 per unit, but at Point C, the poor-quality cost has dropped from $9 to $2.50 per unit. Fifty percent of the reduction in poor-quality cost was labor cost ($6.50 × 50 percent = $3.25). As a result, 20 percent of the value added ($16 ÷ 3.25 = 20 percent), labor force, and equipment were now freed up to produce new products. This increased production per week by 400 units. Two basic assumptions were made in this evaluation: first, that a market was available for the in-

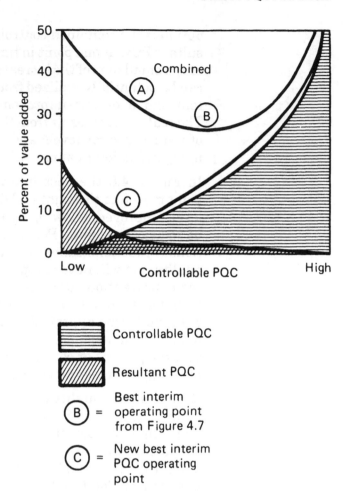

FIGURE 4.8 Poor-quality cost improvement with prevention activities.

TABLE 4.2 Poor-Quality Cost Reduced from Point A to C

	Point A	Point C
Materials cost per unit	$11	$11
Value-added cost per unit	$25	$18.50
Sales price	$42	$42
Profit per unit	$6	$13.50
PQC percentage of value added	35%	9%
Resultant PQC percentage of value added	29%	2%
Weekly production	2000 units	2400 units
Profit per week	$12,000	$32,400
Profit improvement of Point C over Point A		270%

creased production, and second, that the people and equipment freed up when scrap and rework were reduced were the same as those required to increase production capacity. In some cases, neither of these two assumptions is correct.

How Should You Spend Your PQC Dollar?

One of the most frequently asked questions about poor-quality cost is: "What should the PQC be for my company?" There is no good general answer to this question. It varies based on the technology used, where the product is in the development and manufacturing cycle, and the type of product/

service that is being provided. Dana Cound, Vice President of Quality for DiversiTech General, when asked what a company's poor-quality cost should be, gives what he calls his Samuel Gompers answer. (Samuel Gompers was a famous labor leader. When asked what labor wanted, he said, "More.") Cound states, "Whatever your quality costs are, they probably should be less" (10).

As soon as a company starts to embrace a poor-quality-cost system, management starts to ask what percentage of the PQC dollar should be spent on prevention, appraisal, and internal and external error cost to have the system operate at the best interim operating point.

This question does not have a simple answer. The division of poor-quality-cost dollars is highly dependent on the product that is being manufactured, the supporting manufacturing system, where the product is in the product cycle, and its required performance levels. Table 4.3 shows how IBM was spending its poor-quality-cost dollar during the first part of 1980.

James R. Houghton, Chairman of Corning Glass Works, has observed, "We've always been involved with cost reduction and containment of one form or another, but the Quality Management System helps us to focus on cost of quality as we've never done before. We estimate that Corning's cost of quality is approximately 30 percent of sales—and that figure is typical of most U.S. companies" (7).

TABLE 4.3 IBM's Poor-Quality Costs (1980)

	Percentage of revenue dollar	Percentage of total PQC
Prevention cost	2–7	15
Appraisal cost	4–10	25
Resultant PQC	9–23	60
Total	15–40	100

Source: Ref. 12.

Table 4.4 provides a percentage breakdown of poor-quality costs in typical industry groupings. (Because of the difficulty of obtaining data, and because of variations in specific poor-quality-cost systems, the sample shown may not represent the total population.) The data presented in this table is for companies that have been using quality-cost-reporting systems for a number of years and have already implemented parts or all of a quality-improvement process. It also does not include costs related to support groups. In many cases, adding the support-group cost will more than double the total. Table 4.5 provides typical controllable and resultant PQC by industry. If you have not been using poor-quality cost in conjunction with a quality-improvement process, your PQC probably will be much higher.

TABLE 4.4 Poor-Quality-Cost Elements for Major Industrial Groupings by Percentage of Total Poor-Quality Costs

Industry	Prevention	Appraisal	Internal error	External error
1. Accessories	9.9	54.1	32.2	3.8
2. Air frame	13.7	54.6	28.6	3.1
3. Chemicals and applied products	14.1	23.0	42.1	20.8
4. Electronics	21.2	44.6	26.7	7.5
5. Fabrication	10.5	34.4	39.2	15.9
6. Furniture	13.1	35.4	31.4	20.1
7. Instruments	16.6	29.2	24.5	29.7
8. Machinery	11.0	23.6	39.7	25.7
9. Missiles and space	22.5	57.1	15.9	4.5
10. Primary metals	6.5	24.0	48.6	20.9
11. Rubber and plastics	2.7	15.7	64.6	17.0
12. Transportation equipment	8.7	45.2	33.2	12.9
13. Average	12.5	36.7	35.6	15.2

A word of caution in using the data presented in Tables 4.4 and 4.5: The data was collected from many different technical reports and firsthand discussions with quality professionals. The poor-quality-cost elements used in each of the major categories

TABLE 4.5 Poor-Quality Costs by Industry (as Percentage of Sales) for Companies That Have Been Using PQC

Industry	Total	Prevention	Appraisal	Internal error	External error
1. Accessories	6.34	0.63	3.43	2.04	0.24
2. Air frame	4.54	0.62	2.48	1.30	0.14
3. Chemicals and applied products	4.82	0.68	1.11	2.03	1.00
4. Electronics	10.38	2.20	4.63	2.77	0.78
5. Fabrication	4.85	0.51	1.67	1.90	0.77
6. Furniture	2.74	0.36	0.97	0.86	0.55
7. Instruments	7.25	1.20	2.12	1.78	2.15
8. Machinery	4.44	0.49	1.05	1.76	1.14
9. Missiles and space	7.65	1.72	4.37	1.22	0.34
10. Primary metals	6.13	0.40	1.47	2.98	1.28
11. Rubber and plastics	14.70	0.40	2.30	9.50	2.50
12. Transportation equipment	3.89	0.34	1.76	1.29	0.50
13. Average	6.48	0.80	2.28	2.45	0.95

differ from company to company. These differences can cause great variation in the reported costs even if the actual costs were really the same. In all cases the major poor-quality-cost elements (scrap, rework, warranty, and inspection costs) were included.

These figures represent data from 87 companies, and at least three companies are represented in each industry grouping. A list of companies included in this analysis was not supplied due to the proprietary nature of the data. A number of other studies have been conducted on this subject. For example, the Aerospace Industries Association periodically conducts an excellent detailed study called the *Quality Resources Study,* but the distribution of this valuable report is very limited.

What percentage of the sales price should poor-quality cost be? Theoretically, zero. Practically anything over 6 percent of sales (without considering white-collar poor-quality cost) should be of concern to management, and anything under 2 percent of sales should be a cause for rejoicing. Philip B. Crosby's book, *Cutting the Cost of Quality,* uses 4 percent of sales as a guideline, regardless of the product (11).

Impact of Reducing PQC

Reducing PQC is probably a company's best strategy to improve the bottom line. Take a hypothetical company, the James Stevens Foundry, that has the following 10-year projection:

- Constant seven-year net sales of $20 million per year
- Profit = 6 percent of sales
- Year 1 poor-quality cost = 20 percent of sales

If the poor-quality cost could be reduced only 2 percent of sales per year over the next 7 years—reducing the PQC to a target of 6 percent of sales—profit would double in the third year and triple in 6 years (see Table 4.6). The cumulative savings over a 10-year period ($19.6 million) would almost equal the yearly net sales. With that much added profit, your company should be able to pass some of it on to the customer, which should help increase your market share.

TABLE 4.6 Effect of Reduced PQC on Profits (in millions of dollars)

	Years							
	0	1	2	3	4	5	6	7
Sales	20.0	20.0	20.0	20.0	20.0	20.0	20.0	20.0
PQC at 20%	4.0	4.0	4.0	4.0	4.0	4.0	4.0	4.0
New PQC percentage	0	18.0	16.0	14.0	12.0	10.0	8.0	6.0
New PQC $	4.0	3.6	3.2	2.8	2.4	2.0	1.6	1.2
PQC savings	0	0.4	0.8	1.2	1.6	2.0	2.4	2.8
PQC cumulative savings	0	0.4	1.2	4.0	4.0	6.0	8.4	11.2
Old profit (6%)	1.2	1.2	1.2	1.2	1.2	1.2	1.2	1.2
New profit[a]	1.2	1.6	2.0	2.4	2.8	3.2	3.6	4.0
Percent profit change	0	133	167	200	233	237	300	333

[a]New profit is profit at 6 percent of sales plus savings from PQC reductions.

For purposes of clarity, this model has been kept very simple (e.g., holding sales dollars constant). Even in this simplified form, however, it demonstrates the enormous financial payback that quality improvement can bring.

At the 1986 EOQC Conference in Stockholm, Pierre Jaillon and Lemaitre Afnor reported that the cost of failing to meet quality requirements in France during 1981 was estimated to be:

- 150×10^9 francs for industry and the building sector

- 270×10^9 francs for all national industrial activities

I conservatively estimate that the poor-quality cost for all of the United States in 1985 was over 633 billion dollars.

5

Starting a Poor-Quality-Cost System

Being a good manager who wants to reduce cost while improving quality, you are probably champing at the bit to get started with a poor-quality-cost system. This is a crucial point, because the way you start your system will have a major impact on how effective it is and how long it will last. The project must be approached with a well-organized, systematic method. We can divide the implementation phase of a poor-quality-cost system into the following steps:

1. Develop a financial and quality-assurance implementation team.

2. Present the poor-quality-cost concept to top management.

3. Develop an implementation plan.

4. Select a trial area.

5. Start the program in the selected area.

6. Identify and classify poor-quality-cost elements for the selected area.

7. Determine staging for each poor-quality-cost element.

8. Establish inputs to the poor-quality-cost system.

9. Establish required output format.

10. Establish the additional data system required to support the poor-quality-cost system.

11. Review the status with the plant management team.

12. Start the trial period.

13. Review the monthly poor-quality-cost report.

14. Based on findings, modify the program as required.

15. Expand the program to the remainder of the plant and the company.

**Step 1:
Forming the
Implementa-
tion Team**

Many poor-quality-cost systems never get off the ground because they are planned, organized, and implemented solely by the quality assurance (QA) function. Quality assurance traditionally reports defects while it is the controller and the financial areas that report costs, so why not take advantage of these established roles?

This means that the first step in implementing a poor-quality-cost system should be to introduce the concept to the company controller. Show him or her the financial benefits that can be realized by an improvement process based on costs instead of errors. Point out to the controller that he or she is the right person to report the plant's poor-quality-cost data and that this approach will provide a new cost-analysis tool. Quality Assurance should help to set up the data system and sell the concept to the plant management team, but it should be the controller's process.

Putting the poor-quality-cost report in the controller's hands allows it to expand across all functions within the company. This also improves the confidence management will have in the report (cost figures put out by Quality Assurance are often questioned).

What happens if your accounting function is overworked and understaffed, causing the controller to agree with the basic concept but unable to support it with resources? Often this condition occurs in companies that have not implemented a poor-quality-cost system and an improvement process. With poor-quality costs using up to 50 percent of their available staff (a condition that frequently occurs in this type of company), they may not be able to take on another task because they are already working overtime to get out the P & L reports.

Well, all is not lost. It just requires a different strategy. In this case, the quality assurance function should accept the responsibility for pulling together the poor-quality-cost estimate. Once these figures are presented to upper management and the magnitude of the problem is clearly understood, you will gain the support you need from the accounting function. Much of the information is available in the company's present records. All you need to do is to know where to look. The following are rich sources of data that will enable you to prepare an initial poor-quality-cost estimate.

General Ledger   The general ledger is potentially the most valuable source of poor-quality-cost data and, if properly structured, it will provide most of the information you need to compile poor-quality costs for the product. The general ledger was prepared to provide P & L reports and many of the poor-quality costs are deeply embedded in figures that are combined with costs of other activities. For example, the labor hours for a unit include both assembly and test time. The general ledger is maintained by the accounting function, and you will probably have to get some instruction from one of the accountants before it can become an effective, useful source of data for you.

Scrap and   This report provides the data needed for the
Rework Report   manufacturing part of the internal error costs.

Budgets	Often the best way to obtain data related to white-collar and support areas is to look at their budgets. This is a good source of project line-item cost data—for example, approved engineering change costs, training costs, and equipment expenditures. The entire budget for your field service organization should be applied to external error costs.
Operating Statement	This is a good source of revenue data and provides an overview of the total activities.
Capital Equipment List	This is probably the best source to identify appraisal equipment costs and provide you with cost and user department information. On occasion it may become necessary to combine this data with the calibration recall list to obtain the information you need.
Labor Claiming Report	This report is normally broken down by operation to allow you to identify the cost related to appraisal and in-process rework.
Warranty Service Report	Frequently the field service function generates a report that provides warranty cost data. This is a key element in the external error cost summary.

The thrust of this initial estimate should be to accumulate the total poor-quality-cost figure, not to provide an accurate picture of how it is distributed among poor-quality-cost categories. This allows you to make some general assumptions that must be refined later in the program. Some of the items that can be considered elements of poor-quality cost at this point in the program are:

1. All quality assurance and field service budgets.

2. 75 percent of product engineering activities after the design has been released. If the design is released correctly, there is no need for continuing support from Product Engineering unless the customer comes up with a new requirement. Then it should be treated as a new project.

3. 75 percent of the engineering change cost after the design has been released.

4. 75 percent of manufacturing engineering activities after the process has been certified. If the process design is correct, all the manufacturing engineering activity after certification should be directly related to correcting manufacturing problems.

5. Interest on the value of the inventory whose turnover rate is greater than one week, plus the floor space cost required to store this excess inventory.

6. 20 percent of all white-collar personnel costs.

Frequently at this stage of the project, you
will have to rely on educated estimates.
John Heldt, in his self-study guide, *Controlling Quality Costs,* provides the following guidelines (13):

1. Supervisors and line engineers have
an excellent understanding of the
process for which they are responsible. Frequently their own data
sources and understanding of the
process provide excellent insight into
the way the hours and dollars are
being spent. Heldt feels their estimates are accurate to within 2 percent of actual.

2. Percentage allocation of hours and/or
dollars. Heldt states, "If one is reasonably prudent in the estimating
process, the cost of quality allocations should be accurate to within 10
percent."

3. Total poor-quality cost. As a rule of
thumb, Heldt points out that the total direct poor-quality cost will be
surprisingly close to two times the
error cost. (Caution should be applied when using this ratio because it
varies greatly with the type of industry being evaluated.)

It is important always to keep in mind that
your final objective is to have a fully integrated reporting system that is maintained
by the controller's function. With this as a

basic requirement, it is important that you keep the controller informed of your progress. When you have completed your initial study, the data should be reviewed with the controller so that his or her input can be included. It is even a good idea to suggest to the controller that someone from the accounting group present the initial study or at least assist in presenting this data to the company president.

Step 2: Presenting the Concept to Top Management

Once you have gained the support of the controller, you are ready to present the concept of a poor-quality-cost system to upper management. Start as high as you can get—for example, the plant manager or vice president of manufacturing, or the company president. Over and above the obvious advantage of having top management support for the program, you will also find that upper management will grasp the value of a quality report based on cost and profit more quickly than a line manager will—because the language of the board room is dollars whereas the language of the line is defects, yields, and production quantities. If you start at a low level, you may have your plan rejected because your audience does not have the insight to see the real value in measuring and reporting poor-quality costs. Lower management may dismiss it, saying, "It is just another way of cutting the melon, and adds no value." Finance and Quality Assurance should provide an estimate of the company's poor-quality cost based on available data supplemented with conservative estimates.

Make your presentation to upper management crisp and to the point. It should take no more than one hour. You should include examples of poor-quality cost from companies that make similar products, as well as your company's estimated PQC. The proposal should describe how to establish a measure of the major poor-quality expenses, how to track them to show progress in reducing these costs, and how to ensure that management and engineering efforts spent on corrective action will provide maximum return. You should also review your implementation strategy and present additional costs that may be incurred as a result of the implementation program. Allow at least 15 minutes for questions and answers, and (ideally) manager approval. Many people fail in dealing with upper management because they use up all their time without allowing for an opportunity to have the program approved.

Step 3: Developing the Implementation Plan

When establishing an implementation plan, do not try to start with a major revision of the established financial system so you can collect possible PQC elements. Many quality-cost programs have died before they got off the ground because the players were not willing to accept anything less than a complete system; as a result,

they defined a program with major implementation costs that were difficult to justify and thus caused the program to be de-
layed or terminated.

Rule number one in planning a poor-quality-cost system is to start off simply and
then expand the program in stages. A typical five-stage implementation plan would
deal with the various elements in the following order:

Stage 1.   Items that are already avail-
           able in the financial data sys-
           tem.

Stage 2.   Major items that must be in-
           cluded as soon as possible but
           require changes to the finan-
           cial system.

Stage 3.   Detail items that require
           changes to the financial sys-
           tem (by the end of this stage
           you should be reporting a min-
           imum of 80 percent of the total
           blue-collar poor-quality cost).

Stage 4.   Items that require major
           changes in the data-collection
           and -reporting system before
           they can be reported.

Stage 5.   Items needed for a full poor-
           quality-cost picture but that
           have only minor impact on the
           total dollar amounts.

In most cases, Stage 1 items are acceptable
to start the program and provide management with a gross assessment of the potential benefit that can be obtained from the

improvement process. Stage 1 must include scrap, rework, and warranty costs. In some cases, it may be necessary to include in the Stage 1 program some estimates of costs that will not be available until the Stage 2 program is completed, despite the inherent error in estimating cost figures.

Stage 2 costs should be included in the program before you start measuring the results of the improvement process, as there will be a major increase in poor-quality cost as a result of including this additional data in the quality-cost data base. The cost of engineering change activities should be in place for Stage 2.

Allow at least six months between the implementation of Stage 2 and the time Stage 3 is launched. This delay is necessary to allow management time to adjust to and use the data and make it part of their personal measurement plans. It also gives you the chance to build a stable base against which to measure progress before you increase the quality-cost data base. When Stage 3 is implemented, it is good practice to portray graphically both Stage 2 and Stage 3 data for a minimum of six reporting periods to provide a reference base before you drop the Stage 2 data completely.

Stage 4 should be implemented about two years after the start of the program. It is at this stage that many of the white-collar costs are added to the program. This is a difficult stage of the program but a very important one. In many companies, it will more than double the poor-quality cost.

Stage 5 is a nicety that many companies will never reach. It is needed to show that you have considered every part of the poor-quality-cost system, but the additional refinements may not be justified financially. They may also cause annoyance to some of your employees. Don't rush into this stage until the system is firmly established. It should start no sooner than five years after the beginning of the program, and even then you may want to include only selected items.

Remember that this is not a financial report that will be audited. It is a trend and magnitude indicator. Its purpose is to define major problem areas—to sift the wheat from the chaff. Its purpose is to get management's attention, so don't try to make it 100 percent accurate; in fact, a good way to kill the program is to insist on an absolutely accurate data base. Don't try to get the pennies to cross-balance. This puts unjustifiable cost and delay in the system, not to mention the additional bureaucracy that is required. A report that is 80 percent accurate may serve the purpose as long as it is constant and covers the major activities and costs. With PQC reporting, consistency is the most important factor.

Step 4: Selecting a Trial Area

The next step is to select a trial area. It may be one plant in the company structure or one product line in a plant. The keys are:

1. It should be self-contained. This allows the total cost impact to be eval-

uated and compared to a value-added
or sales-cost base.

2. It should already have a good cost
    data base. The more detailed the
    present data system, the less time
    and cost will be required to install
    the trial PQC system.

3. It should be an area where manage-
    ment is open to new ideas. The man-
    agers in the area should be willing to
    accept change as a challenge, not as
    an obstacle to overcome.

4. It should be an area that needs to
    improve its quality and that will de-
    rive long-term benefits from improve-
    ment. This is necessary so that you
    will have management and engineer-
    ing support for implementing the
    quality-improvement program and
    can present meaningful success
    stories to upper management at the
    completion of the trial period.

**Step 5:
Starting the
Program**

Once you have selected the trial area, present the program to the management of
that area. The poor-quality-cost system
should be presented as a management tool,
not a quality or financial program. It exists
to help manage the product effectively, improve productivity, reduce cost, and allow
management to provide products and services that better meet the customer's expectations without suboptimizing the system.
The operating unit manager must be involved as early as possible in the planning
cycle, and should be given the opportunity

to contribute to the many trade-off decisions that are required in starting the program.

Step 6: Identifying and Classifying the Cost Elements

The next step is to identify all the PQC cost elements related to the trial area. This list will include both poor-quality cost and operating-cost elements. Once the cost-element list is completed, each element should be classified in one of the six categories listed in Figure 5.1. You will note that an individual element can be recorded in more than one category by using percentages of the total cost.

When you have completed the listing and classifying, go back to the basic definitions

Department cost elements	Operating costs	Prevention costs	Appraisal costs	Internal error costs	External error costs	Equipment costs
Assembly hours*	90%		10%			
Operator training		100%				
Management training		100%				
Building rental	50%	5%	25%	20%		
Scrap and rework				100%		
Inventory costs	60%			20%	20%	
Final test			100%			
Test equip...						

FIGURE 5.1 Classification of department cost elements.

of the quality-cost elements and see if any have been left off the list. (The items left off the list frequently fall into the categories of external error cost and prevention cost.) Then complete the matrix.

Step 7: Staging Each PQC Element

Using the complete cost-element-list matrix as a base, you are now ready to determine when each quality-cost element will be introduced into the system. Figures 5.2,

	Implementation Stages				
Activities	1	2	3	4	5
I. Prevention					
• quality planning			x		
• training of manufacturing operators and quality inspectors		x			
• process control planning		x			
• quality data collection and analysis system planning			x		
• quality reporting equipment costs				x	
• preventive action					x
• procedure preparation		x			
• quality motivation programs				x	
• other function quality plans			x		
• test equipment planning					x
• product engineering pre-customer ship evaluation				x	
• vendor qualification		x			
• QA early entry	x				
• training for equipment maintenance		x			

FIGURE 5.2 Setting priorities for prevention poor-quality-cost elements.

5.3, 5.4, and 5.5 provide examples of how this is accomplished. Three factors must be considered:

1. The present cost data system

2. The importance and magnitude of the poor-quality-cost element

3. The cost and/or impact of changing the present cost-data system to collect the required data

	Implementation Stages				
Activities	1	2	3	4	5
II. Appraisal					
• quality audits	x				
• manufacturing inspection and testing	x				
• inspection of purchased items		x			
• outside endorsements					x
• maintenance of inspection and test equipment		x			
• quality analysis		x			
• field performance testing		x			
• measurement and QA data processing			x		
• in-process control charts			x		
• set-ups for inspections and tests				x	
• installation check-outs	x				
• packaging evaluations				x	
• assurance testing before first customer shipment			x		
• field data systems		x			
• quality data collection and analysis operations		x			
• installation testing	x				

FIGURE 5.3   Setting priorities for appraisal poor-quality-cost elements.

Activities	Implementation Stages				
	1	2	3	4	5
III. Internal Error					
• scrap and rework	x				
• troubleshooting and repair	x				
• retest	x				
• Materials Review Board activities				x	
• failure analysis		x			
• added inventory costs					x
• downgrading				x	
• reinspecting rejected supplier lots			x		
• corrective actions		x			
– process changes		x			
– engineering changes		x			
– retraining			x		
– rewriting documents				x	
• screening bad from good items	x				
• productivity loss				x	

FIGURE 5.4 Setting priorities for internal error poor-quality-cost elements.

Activities	Implementation Stages				
	1	2	3	4	5
IV. External Error					
• product liability suits				x	
• complaint handling					x
• field troubleshoot, repair, and retest	x				
• field repairman training				x	
• return parts handling and repairs			x		
• failure analysis			x		
• engineering changes to repair field problems		x			
• field corrective action			x		
• repair manuals		x			
• marketing errors				x	
V. Equipment					
• cost to build, debug, and install			x		
• cost of purchased equipment and installation			x		

FIGURE 5.5 Setting priorities for external error and equipment poor-quality-cost elements.

**Step 8:
Establishing
Input to the
System**

During the classification process, you will define what elements need to be included during Stage 1 of the poor-quality-cost system and the additional data that is needed for Stage 2. At this time, each of the elements planned for Stage 2 should be evaluated to determine how the required data will be obtained and the cost involved in obtaining it. The result will be a program plan to implement Stage 2. In some cases, depending on the adequacy of Stage 1 base data, it may be necessary to implement Stage 2 in the trial area before you start reporting poor-quality cost. But in most cases, Stage 2 should be implemented when the program is expanded to the rest of the plant.

**Step 9:
Establishing
Output
Formats**

The next step is to establish the output formats best suited to your operation. This step may be complex because the poor-quality cost data must be presented in different ways, depending on the level of management reviewing the report and how they should use the data. Company upper management wants to see data presented by division, plant, and product, whereas line management needs the PQC data by department.

Corporate
Level Reports

Typical output formats that can be used for the corporate-level reports are shown in Figures 5.6 and 5.7. Figure 5.6 provides a view of the total plant's product poor-quality cost as a percentage of value added. The left side provides a long-range view of the quality-improvement progress and the right side shows the monthly variation. Depending on the way your company's upper management reacts to the data, it may be advantageous to show only the left side of the report so they do not overreact to normal monthly variations. If this is done, report the data on the left side on a quarterly basis instead of the six-month basis used. The graph also provides a best-fit trend line to keep management from trying to draw the line by eye. Figure 5.7 presents poor-quality cost per month in millions of dollars.

Figures 5.6 and 5.7 use the same data but present it in two different ways, leading to very different conclusions. Figure 5.6 leads you to believe that the quality-improvement process is performing well, and, in fact, it is. Figure 5.7 is misleading because it does not use a constant data base. The information that is not taken into consideration in the way Figure 5.7 presents the data is that the plant is now producing more product than it was in 1982 (increased production schedules).

In fact, both graphs need to be presented because Figure 5.7 shows that there is still a lot of gold to be mined in "them thar hills." And total PQC dollar value always gets management's attention. Figure 5.8 provides the data needed to answer any questions that corporate management might

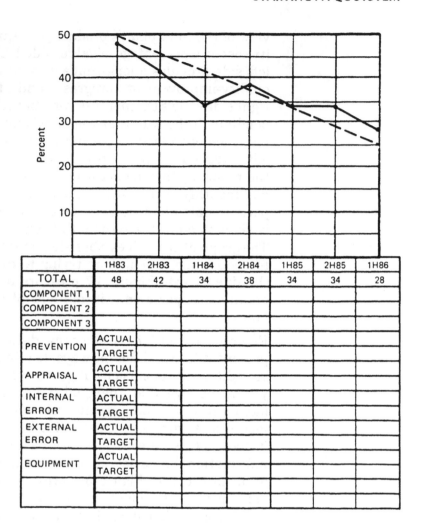

	1H83	2H83	1H84	2H84	1H85	2H85	1H86
TOTAL	48	42	34	38	34	34	28
COMPONENT 1							
COMPONENT 2							
COMPONENT 3							
PREVENTION ACTUAL							
PREVENTION TARGET							
APPRAISAL ACTUAL							
APPRAISAL TARGET							
INTERNAL ERROR ACTUAL							
INTERNAL ERROR TARGET							
EXTERNAL ERROR ACTUAL							
EXTERNAL ERROR TARGET							
EQUIPMENT ACTUAL							
EQUIPMENT TARGET							

FIGURE 5.6 Value-added poor-quality-cost report for top management.

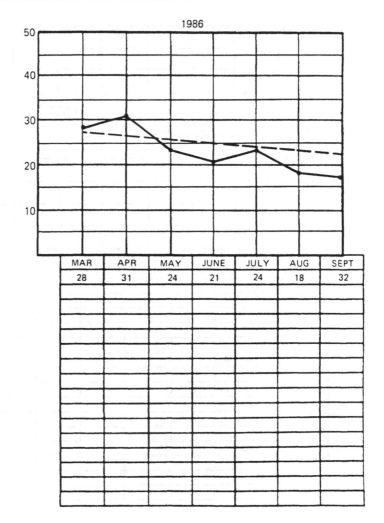

	MAR	APR	MAY	JUNE	JULY	AUG	SEPT
	28	31	24	21	24	18	32

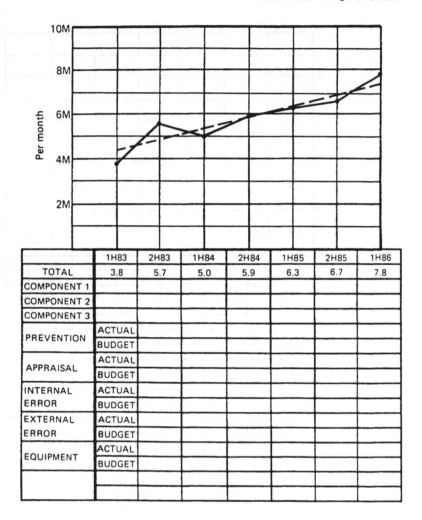

	1H83	2H83	1H84	2H84	1H85	2H85	1H86
TOTAL	3.8	5.7	5.0	5.9	6.3	6.7	7.8
COMPONENT 1							
COMPONENT 2							
COMPONENT 3							
PREVENTION ACTUAL							
PREVENTION BUDGET							
APPRAISAL ACTUAL							
APPRAISAL BUDGET							
INTERNAL ERROR ACTUAL							
INTERNAL ERROR BUDGET							
EXTERNAL ERROR ACTUAL							
EXTERNAL ERROR BUDGET							
EQUIPMENT ACTUAL							
EQUIPMENT BUDGET							

FIGURE 5.7   Total dollar poor-quality-cost report for top management.

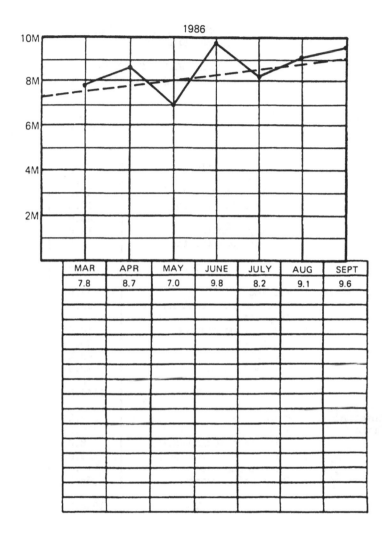

	MAR	APR	MAY	JUNE	JULY	AUG	SEPT
	7.8	8.7	7.0	9.8	8.2	9.1	9.6

JOHNSON PLASTICS COMPANY
San Francisco Plant

POOR-QUALITY COST DATA
(Second Quarter 198X)

ELEMENT	ACTIVITY	PQC (K $)	% VALUE ADDED	% PQC
Prevention	Design evaluations	22.0	2.8	10
	Training	46.0		
	Quality motivation	58.0		
	Reliability testing	112.5		
	Early entry	48.0		
	Quality planning	12.0		
	Process qualification	83.0		
	Total	381.5		
Appraisal	Testing	485.0	8.4	30
	Environmental testing	22.0		
	Inspection	526.0		
	Engineering studies	10.0		
	Calibration	36.3		
	Audits	43.2		
	Stress testing	22.0		
	Total	1144.5		
Internal Errors	Rework	525.0	11.2	40
	Scrap	785.0		
	Corrective Action (QA)	8.0		
	Corrective action (ME)	29.0		
	Process changes	115.0		
	Engineering changes	64.0		
	Total	1526.0		
External Errors	Warranty	512.0	5.6	20
	Field service training	18.0		
	Parts stocking	60.5		
	Failure analysis	10.5		
	Engineering changes	162.0		
	Total	763.0		
	Grand total	3815.0	28.0	100

FIGURE 5.8   Detail data for top management.

ask about how the poor-quality-cost dollar is being spent.

Plant Management Reports

The plant management team uses the poor-quality-cost information in a completely different way than top management does, because they need it to direct the details of the quality-improvement process. This need requires them to have a picture of how each function within the plant and each of the major products is performing, and a breakdown by major poor-quality-cost element (controllable and resultant PQC).

Poor-quality-cost performance by product and function can be graphically displayed using the same format shown in Figures 5.6 and 5.7 because each function has a value-added base for its activities. For example, the value-added base for the financial function is equal to all the costs incurred directly and indirectly by that function (salary, benefits, equipment, materials, training, etc.). For the purposes of a poor-quality-cost system, let's accept value added as being equal to the cost of performing the activity. Figure 5.9 shows another way of graphically presenting the same data. You will note that this format has the advantage of presenting the total cost picture on one page. By carefully examining these graphs, you can see that management strategy for reducing poor-quality cost is one of increasing preventive cost to reduce the cost of the other three poor-quality-cost elements. In addition, these graphs show

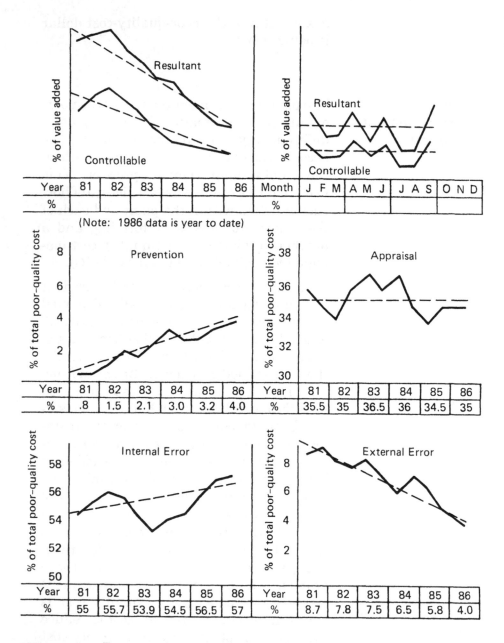

Year	81	82	83	84	85	86	Month	J	F	M	A	M	J	J	A	S	O	N	D
%							%												

(Note: 1986 data is year to date)

Year	81	82	83	84	85	86	Year	81	82	83	84	85	86
%	.8	1.5	2.1	3.0	3.2	4.0	%	35.5	35	36.5	36	34.5	35

Year	81	82	83	84	85	86	Year	81	82	83	84	85	86
%	55	55.7	53.9	54.5	56.5	57	%	8.7	7.8	7.5	6.5	5.8	4.0

FIGURE 5.9   Poor-quality-cost element report.

that management is placing emphasis on finding the problems within the plant, because the percentage of poor-quality cost spent for internal error cost and appraisal cost remains flat, while external error costs decrease significantly.

The same basic format can be used to break down data for each of the products and functional areas into more detailed reports needed at the project and line-management levels.

Analyzing PQC Reports

Using the data presented in Figure 5.8, the following conclusions can be made:

1. The ratio between PQC categories is good but the company should not be spending 28 percent of the value-added content on PQC. It should be in the 6 to 10 percent range.

2. The appraisal activities are effective because there is a two-to-one ratio between internal and external errors; there is no need to change the ratio at this time.

3. Almost 50 percent of the PQC is the result of rework, scrap, and warranty costs. The first area that needs attention is reducing internal and external errors.

4. With $226,000 of the poor-quality cost devoured quarterly by engineering changes, effort should be expended to understand why these problems were not corrected before

the product was released to the customer.

The
Improvement
Team

The improvement team requires much more detailed data to allow them to attack the correct problems. This data can be supplied to them in a computer list sorted by cost of major problems. Frequently the major problems are identified by the improvement team, and the PQC system is set up to provide them with the cost data necessary to set priorities and to track the progress in reducing the PQC related to the priority problems.

Step 10: De-
fining Addi-
tional Data-
Assistance
Requirements

Now that you have defined the required outputs and understand the currently available input data, it is necessary to define what *additional* input data is required before a meaningful poor-quality-cost evaluation can be made to support the trial run in the selected area. If this minimum data is not available, then key input elements from Stage 2 must be completed before you enter the next phase. The implementation team should now involve the information-services people in the planning cycle so they can understand the concept and estimate the programming cost required to support each stage of the poor-quality-cost system.

You are also at the point where the implementation team should develop a plan to set up a system that will collect the input data needed to implement Stages 2 and 3. This activity will go on in parallel with a trial evaluation, and should be scheduled to be completed so that all the Stage 2 activities can be implemented as soon as the trial period is completed in the selected area. During the trial evaluation, much of the data reporting can be done by hand, but this is not practical for the large quantities of information available when the program is implemented on a plant-wide or company-wide basis.

Step 11: Reviewing Status with Plant Management

Now you are ready for the second review with plant management. The purpose of this review is to reacquaint the managers with the poor-quality-cost system, to provide them with detailed understanding of the implementation plan, and to show them the output formats that will be used. Probably the best time to do this is when you have your first monthly report from the trial area. Take time to explain the report in detail and show how the report should be used. Although you are only at Stage 1, take time to estimate what the total poor-quality cost is for the product.

This is an extremely important meeting. It will probably be the first time that your management has seen the magnitude of their company's quality problem, and many of them will be shocked to see how much money is being wasted. Many won't believe it, and others will try to reason it away by saying that the data is not typical of the whole company. It is one thing to have someone estimate that your company could be putting as much as 25 percent of the company's expenditures into appraisal and error activities, but it is quite another thing to be shown that their plant is performing that poorly.

Before this meeting, they were told about the *other* person's problems, the companies that must have had poor management to be that badly off, but now you are talking about *their* plant and *their* company. Now it's a completely different matter. Before, they told themselves, "Yes, I understand, but that doesn't apply to my company. We manage our business much better than that."

This first report will be a shocker in most cases. Once they begin to realize that they can increase their profits by as much as 80 percent if they do it right every time, they will begin pressuring you to implement the poor-quality-cost system throughout the plant. Resist the pressure. Complete your trial run, automate your reporting system, and complete the work required to bring Stage 2 on line before you expand.

For the trial area, the best way to decrease poor-quality cost is to concentrate on reducing high-cost error items using the methods described in *The Improvement Process* (see "Suggested Additional Reading," page 189). This includes upgrading the training programs to ensure that each individual assigned to the program has the knowledge and ability to do the job right every time. Now that management knows what poor quality is costing, they have the justification required to invest engineering and management effort to improve the output quality. Obtaining management's commitment to support the improvement process is critical to the future success of the poor-quality-cost system. Management should assign one manager to be responsible for driving down PQC in the trial area.

Step 12: Starting the Trial Period

The trial should last a set time period, and all support areas should have the necessary human resources identified and committed to the program for the trial period. The first thing that should be done as you start the trial period is to form department- and system-improvement teams. The system team is responsible for reviewing the poor-quality-cost data and determining how the controllable elements of poor-quality cost should be spent. A meeting chaired by the manager responsible for the trial area should be held once a week and action logs maintained. The members of the system-improvement team should be measured on reduction in poor-quality cost.

Step 13: Reviewing the Monthly Report

Once a month, a poor-quality-cost report should be published, along with a brief summary of the prevention activities implemented by the improvement teams. All activities should include an estimate of projected savings and improved quality. The report should be reviewed in detail with the plant management team, by finance, and by the improvement teams. Other functional managers should be periodically invited to sit in on the monthly meetings so they will become acquainted with the poor-quality-cost system and be more willing to support the process as it expands to their own areas.

Step 14: Modifying the Program Based on Experience

As you proceed through the trial period, you will find that changes are needed to make the poor-quality-cost system uniquely suited to your company's operation. After you are six months into the trial period, start a public-relations project to gain understanding and input from other functions or plants where a poor-quality-cost system will have a favorable impact. Try to modify the system so that it meets everyone's needs without making it so complex that it is unmanageable. Your system at this point should be a living, changing process that is modified based on experience and on your customer's (management's) expectations.

Step 15: Expanding the Program

At the end of the trial period, the management team should have the necessary data to determine if the program should be expanded or terminated. If all involved did their jobs, then the decision should be to expand the system throughout the plant and the company. This decision means that a company implementation strategy will have to be developed. It is relatively simple to expand the poor-quality-cost system from one area to the total plant and to the company because most plants use common accounting systems. When you talk about expanding the program on what could be a worldwide basis, it becomes more complex, of course.

A Word of Caution

A poor-quality-cost system is designed to highlight major financial opportunities for improvement and to show improvement trends. This system is not designed to be 100 percent accurate or to be used to compare product lines, plants, or companies. Upper management should resist the urge to question why there is a difference in poor-quality cost between areas and concentrate on ensuring that the trend is acceptable.

Unfortunately, no matter what the financial or quality communities tell upper management about not using the data to compare plants and products, this advice normally falls on deaf ears. When you see two graphs in a row and one plant's poor-quality cost is running at 11 percent of value added and the next plant's poor-quality cost is running at 27 percent of value added, there is an irresistible urge to ask *why*. For this very reason, detailed PQC

TABLE 5.1   Detailed PQC Definitions for Manufacturing
Process Change

Category
  Manufacturing process change equipment qualification other
  than design change or engineering change (includes equipment
  requalification).

Internal error

General description
  This category covers line or process changes that don't require
  engineering changes. Principal examples are consolidation of
  subassembly/final assembly operations for cost-reduction purposes, changing the sequence of operations for ease of manufac-
  turability, thus reducing chance of error or improving quality of
  the produced product. In semiconductor manufacture, this category also includes the resetting of equipment and instruction so
  that existing documented processes and specifications are fol-
  lowed to achieve output within targeted distribution ranges.
      Equipment requalification covers normal evaluation to deter-
      mine if the equipment meets specifications to support the ac-
      ceptance or rejection of product lots based on established
      criteria.
      OSHA changes due to ground rules should not be identified as
      a manufacturing engineering failure cost. Also, those
      changes as a result of a tech update or invention should be
      recognized as cost of doing business.
      Commencement/termination points
        Commencement: start of qualified production
        Termination: end of production

Participating functions and tasks

Manufacturing engi- neering	Evaluate alternatives, investigate vari- ances. Revise operations, routings, equipment settings; provide re- instruction.
Quality assurance	Establish acceptance position on con- formance of changes to meet specifi- cations.

TABLE 5.1 *Continued*

Materials management	Revise material flow and related controls.
Facilities services	Provide equipment/facility support.
Development/ product engineering	Assist manufacturing engineering, as required.
Manufacturing	Retrain operators; provide operator assistance in collecting evaluation data.

definition sheets should be prepared and used (see Tables 5.1 and 5.2). These sheets provide specific ground rules for the PQC elements at different locations. This detail is necessary to assure that each plant implements poor-quality-cost systems using the same elements in the same way. This type of detailed description minimizes the inconsistencies between plants and eliminates a lot of hard-to-answer management questions.

To control the implementation of the company's poor-quality-cost system, you need to repeat Steps 1–3 and 6–10, described in this chapter.

As you implement the company poor-quality-cost system, quarterly poor-quality-cost reviews should be held with the company president and vice presidents. The total company poor-quality-cost picture and related major corrective and preventive activities should be presented at these meetings.

TABLE 5.2 Detailed PQC Definitions for Development and
Qualification of Inspection Procedures

Category
  Development and qualification of inspection procedures, design
  and application of functional gauges, testers, etc.

Appraisal

General description
  Provide QA and manufacturing with nonstandard gauging or inspection equipment and procedures to perform an accurate, re-
  peatable measurement of a process parameter.

Commencement/termination points
  Commencement: product release
  Termination: end of product life

Participating functions and tasks

Manufacturing engineering	Develop design and operating instructions.
Quality assurance	Qualification.
Manufacturing	Develop training package.
Programming	Develop software.
Facilities services	Produce the equipment.

6

White-Collar Poor-Quality Costs

John F. Akers, President of IBM, stated on March 13, 1984, "Our studies show that more than 50 percent of the total cost of billing relates to preventing, catching, or fixing errors" (4).

For years the accounting and quality assurance functions have measured and tried to control the process that produces the products that will be delivered to your customers. As a result, great strides have been made in reducing direct product cost while overhead costs have continued to increase at a rate that exceeds the reduction in direct costs. In most companies white-collar poor-quality cost runs 20 to 40 percent of the white-collar area's total budget. This means that you need to expand your quality-improvement activities from the manufacturing areas if your company is going to optimize its profits. To help put a stop to this runaway overhead cost, it is important that the PQC include white-collar PQC.

Companies such as Westinghouse and IBM are making white-collar PQC part of their total poor-quality-cost system. J. R. Forys, Manager of the Reliability and Quality Control Department at Westinghouse Electric, when discussing their PQC system, reported (14):

> The costs of quality, however, are much more than scrap, rework, and warranty expenses; we now also measure and report departmental (i.e., engineering, marketing, etc.) white-collar lost time. Balance-sheet costs such as interest lost on uncollected receivables, interest lost on invoicing delays, and excess inventory carrying costs are also being tracked and reported.

White-Collar Errors

Philip B. Crosby, when he was Vice President of IT&T, stated, "The pen is messier than the soldering iron" (11). How very true. Today the white-collar worker accounts for more of our nation's poor-quality cost than the blue-collar worker. As obvious as all this sounds, it is very difficult to develop a system to report white-collar PQC because today's accounting systems have never focused on the detailed operation in the white-collar area. As a result, only some very gross figures can be obtained from the accounting record, for example:

- Cost of engineering changes
- Cost of program slippage

- Product engineering cost after design release

- Manufacturing engineering cost after the process is qualified

- Added cost because bills were not paid on time

- Absenteeism in excess of 2 percent

- Outside education to improve performance

But these poor-quality costs are only the tip of the iceberg. Appraisal and error costs in the white-collar areas are growing every day in your company. Typical PQC are those of:

- People waiting to go into a meeting room because the previous meeting did not end on schedule

- Meetings starting late because people did not arrive on schedule

- Delays because equipment is down (e.g., secretaries going to another building to make copies because the copier in their building is down)

- Recording errors (e.g., putting a decimal point in the wrong place)

- Communication errors (e.g., management not providing detailed instructions and ensuring that the employee understands what the manager wanted)

- Upper management signoff (e.g., two levels or more of management approval required because the first-level manager cannot be depended on to make the right decision)

- Inspection and return of supplied material (e.g., applied to such things as stationery supplies, medicine for the medical department, clean-room garments, development laboratory components, etc.)

- In-house training for white-collar employees, in both on-the-job and formal classes

- Dismissal of employees due to poor performance

- Nonbusiness-related phone calls

- Missed commitments that allow problems to continue or keep other people from completing their assignments

Understanding White-Collar PQC

In an effort to reduce these types of undesirable costs, some of the techniques that have been proven effective in the manufacturing environment are now being applied to the white-collar areas. In many companies the business process is being viewed in much the same way as the product process. Table 6.1 lists typical business processes where manufacturing process control methods can be applied effectively. In most companies standard process-control theory can be applied to more than 75 percent of the business process. To accomplish this the following steps should be taken:

TABLE 6.1 Typical Business Processes

Personnel	Employee relations
	Executive resources
	Personnel resources
Production control	Physical inventory management
	Disbursements
	Parts planning and ordering
	Consigned materials control
Industrial engineering	Cost estimating
	Process planning
	Space utilization
Quality assurance	Supplier qualification
	New product qualification
	Field reporting
Product engineering	Engineering change control
	Services cost estimating
Finance	Cash control
	Accounts receivable
	Ledger control

1. Define the critical business processes.

2. Assign one person the responsibility for each critical business process.

3. Establish the beginning and end points for the process.

4. Determine who the customers are, and understand and document their expectations.

5. Flow-diagram the process.

TABLE 6.2   Typical White-Collar Measurements

Process equipment downtime

Number of engineering changes per part number

Defects per K lines of code

Forecast versus demand

Revenue versus plan

Percentage of meetings that start on schedule

Number of missed schedules

Inventory turnover rate

Percentage of items received but not delivered in 24 hours

Orders not filled within 24 hours

Errors found during design review

Percentage of suggestions answered within two weeks

Percentage of variances from budget

Number of reviews before plan is approved

Percentage of letters retyped

Percentage of letters dictated

Number of security violations

Personnel turnover rates

Number of complaints per month

Percentage of purchase orders changed

Interest lost on uncollected receivables

Interest lost on invoicing delays

Excess inventory carrying costs

---

6. Define what inputs are required and develop specifications for each.

7. Establish appropriate measurements at process control points. (Table 6.2 lists typical white-collar measurements.)

8. Develop feedback loops that provide individuals with feedback on their performance.

9. Implement an improvement process.

(Additional information about applying process controls to the business process is available in *The Improvement Process,* see "Suggested Additional Reading," page 189.)

Typical PQC by Function

The following list was prepared to help you understand poor-quality cost in the white-collar functions.

1. Controller's PQC
   A. Controllable PQC (prevention and appraisal)
      timecard reviews
      capital equipment reviews
      financial report reviews
      invoicing reviews
   B. Resultant PQC
      billing errors
      incorrect accounting entries
      payroll errors
2. Software PQC

    A.   Controllable PQC (prevention and appraisal cost)
code verification
review of system specs
performance testing

    B.   Resultant PQC
system failure
program restarts
output errors

3. Plant administration PQC

    A.   Controllable PQC (prevention and appraisal cost)
security
facility inspection and testing
machine maintenance training

    B.   Resultant PQC
disclosure of corporate secrets
facilities redesign
incorrect labor levels
equipment downtime
idle equipment

4. Purchasing PQC

    A.   Controllable PQC (prevention and appraisal cost)
vendor reviews
periodic vendor surveys
follow-up on delivery dates
strike built-in costs

    B.   Resultant PQC
line-down cost
excessive inventory due to suppliers
premium freight cost

5. Marketing PQC

    A.   Controllable PQC (prevention and appraisal cost)

               sales material review
               marketing forecast
               customer surveys
               training of sales personnel

    B. Resultant PQC
         overstock
         loss of market share
         incorrect order entry

6. Personnel PQC
    A. Controllable PQC (prevention and
         appraisal cost)
         prescreening applications
         appraisal reviews
         exit interviews
         attendance tracking
    B. Resultant PQC
         absence rate
         turnover rate
         cost of filling positions of people
          who left the company
         grievances

7. Industrial engineering PQC
    A. Controllable PQC (prevention and
         appraisal cost)
         packaging evaluations
         layout reviews
         OSHA reports
         inspection of contract work
    B. Resultant PQC
         OSHA fines
         shipping damage
         redo of layout
         paying contractors for poor work
         increased handling cost due to poor
          layout

8. Information systems PQC
    A. Controllable PQC (prevention and
        appraisal cost)
        cost benefit analysis
        verification of input data
        backup storage
    B. Resultant PQC
        rerun cost
        overtime
        program updates
        cost of action taken based on faulty
            information
        parallel computer system

Collection of White-Collar PQC Data

As much as 50 percent of the white-collar work effort is lost to poor-quality cost. Douglas D. Danforth, Chairman of Westinghouse Electric, reported, "Over 60 percent of our employees today are 'white-collar,' information-oriented workers. And white-collar costs are over half of all our quality costs" (15). With a problem as big as this, it is obvious that something has to be done to provide management with the numerical assessment of where the company is and provide them with a way to measure what progress is being made to reduce white-collar poor-quality cost. In most cases, expanding the accounting system to penetrate into the white-collar areas is not the best way to collect PQC data, due to the excessive costs of gathering the required data and the impact on the white-collar employees' morale. As a result, the most practical approach is either random time sampling (work sampling) or self-analysis.

Random Time Sampling

Random time sampling involves the use of an expert auditor who randomly samples the activities of the people and the status of the equipment in the area that is being observed. This is a very strong tool and can yield a great deal of valuable data that applies to more than just poor-quality cost. It offers the following advantages over self-recorded data:

- More accurate data
- Less variation in data
- Less employee distraction
- Reduced employee paperwork
- Less biased reporting
- Reduced time required to analyze data

Some of the disadvantages of random time sampling are:

- Requires a well-trained auditor.
- Sometimes impressions do not reflect actual work status.
- Employees often react differently when the auditor is in the area, providing data that is not representative of the normal process.

This procedure is well known and well documented. It is based on work done by

L. H. C. Tippett in the 1930s. For more information on this subject, read *Work Sampling* by R. E. Heiland and W. J. Richardson (see "Suggested Additional Reading," page 189).

Self-Analysis

Two self-analysis methods have proven effective in poor-quality-cost analysis: personal logs and department activity analysis. To apply the personal-log technique to collect PQC data, target departments are selected that represent the activities in a major function. At department meetings, PQC is explained and the employees are instructed in how to maintain a personal log in which their activities are recorded and categorized into five classifications: basic work, prevention, appraisal, internal error, and external error activities. When a single activity includes more than one of the five classifications, the individuals are instructed to divide the time between the classifications. The employees should be encouraged to record comments related to why the errors occurred and what can be done to eliminate them. In some departments it is most effective to prepare a list of activities that the department is involved in; the employees then just log time to the appropriate categories, which reduces recording and analysis time.

The personal log should be used to sample the department's activities, so it is necessary to maintain the log only for a short period of time—one week to a month. It may not be necessary for all the members of the department to maintain personal logs if many of them are doing the same activities. The actual duration and sample size must be based on the individual department activity and its activity cycle.

The department activity analysis is another very effective way of accumulating poor-quality cost data for the white-collar areas. It also provides an excellent tool that will help the department improve the quality of its output. During the department activity analysis, the employees define the department's mission and its major work activities (see Figure 6.1). They then define who the customer is for each activity. Once the customer is defined, meetings are held with the customer (normally a department within the company) to develop specifications for the output to reflect the customer's expectation and to set the minimum performance standard for the department. The customer signs off on this form, indicating agreement with the output specifications (see Figure 6.2).

The department employees then analyze the activity, to estimate how much time is devoted to it, and subdivide the time according to the five subcategories: basic work, prevention, appraisal, internal error, and external error (see Figure 6.3). The last phase of the department activity analysis is to define what inputs are required to perform the activity successfully, where the department gets these inputs from, what the quality expectations are for this input

DEPARTMENT ACTIVITY ANALYSIS

Function Name

Department Name   Dept. No.

DEPARTMENT MISSION:

LIST MAJOR OR ALL ACTIVITIES/TASKS/RESPONSIBILITIES OF DEPARTMENT:

Manager's/Preparer's Name   Date   Extension

Department Approval:

_____   _____   _____
_____   _____   _____
_____   _____   _____
_____   _____   _____
_____   _____   _____

FIGURE 6.1 Department activity analysis: mission/major activities.

NOTE: USE ADDITIONAL PAGES IF MORE SPACE IS NEEDED.

ACTIVITY:	DEPT.	DATE	PREPARED BY:

OUTPUT

WHAT ARE THE OUTPUT REQUIREMENTS THAT YOU AND YOUR CUSTOMER HAVE AGREED TO?
o
o
o
o
o
o

WHAT ARE THE QUALITY MEASUREMENTS THAT WILL SHOW IF YOUR OUTPUT MEETS
REQUIREMENTS, AND HOW WILL THEY BE MEASURED?
o
o
o
o
o

Customer Approval:

FIGURE 6.2 Department activity analysis: customer requirements.

NOTE: USE ADDITIONAL PAGES IF MORE SPACE IS NEEDED.

ACTIVITY:	DEPT.	DATE	PREPARED BY:

VALUE ADDED -- WORK ACCOMPLISHED IN DEPT.

WHY DO:

VALUE ADDED:

IMPACT IF NOT DONE:

HOW MANY HOURS/WEEK ARE SPENT ON THIS ACTIVITY? _____ HRS/WK

THESE HOURS ARE EITHER A COST OF DOING BUSINESS, A POOR QUALITY COST (PQC),
OR SOME COMBINATION OF THE TWO.   HOW DO YOU CLASSIFY THEM?

        BUSINESS   _____   HRS/WK
        PQC   _____   HRS/WK

THOSE HOURS THAT ARE PQC CAN BE FURTHER CLASSIFIED INTO PREVENTION,
APPRAISAL, AND FAILURE.   WHAT ARE THEY?

        PREVENTION _____   HRS/WK
        APPRAISAL _____   HRS/WK
        FAILURE   _____   HRS/WK
        TOTAL PQC _____   HRS/WK

FIGURE 6.3   Department activity analysis: value-added analysis.

NOTE: USE ADDITIONAL PAGES IF MORE SPACE IS NEEDED.

ACTIVITY:	DEPT.	DATE	PREPARED BY:

INPUT

WHAT:

FROM:

WHAT ARE THE INPUT REQUIREMENTS THAT YOU AND YOUR SUPPLIER HAVE AGREED TO?
o
o
o
o
o

WHAT QUALITY MEASUREMENTS WILL SHOW IF SUPPLIER'S OUTPUT MEETS REQUIREMENTS?
o
o
o
o
o

Supplier Approval:

_____   _____   _____
_____   _____   _____
_____   _____   _____
_____   _____   _____
_____   _____   _____

FIGURE 6.4 Department activity analysis: supplier analysis.

(input specifications), and how the depart-ment will provide feedback to the suppliers on the quality of the input. The supplying area signs off on this analysis, indicating that they understand and agree with the analysis and the supplier requirements (see Figure 6.4). As you can see, the depart-ment activity analysis provides much more than just poor-quality-cost data. It also es-tablishes the required supplier-customer relations needed to direct the improvement process in the white-collar area.

Because both random time sampling and self-analysis techniques are used primarily to report status and to measure improve-ment in the white-collar areas, the evalua-tion should be repeated about once a year or at the end of a specific improvement cycle.

John Akers, when addressing white-collar PQC, said, "We use special surveys, sam-ples, and site management reviews to gather cost of quality information that may not be visible in our current income and expense statements" (7).

Service Industries' PQC

A poor-quality-cost system is an important part of such service-industry areas as bank-ing, health services, traffic, public utilities, and government. In each case, controllable and resultant PQC is high, and measure-ment is imperative to obtain the proper focus. For example, delinquent taxes run in the billions of dollars every year. An audit of the Aid to Families with Dependent Children in the Washington, D.C. area indicated that 55 percent of the households receiving aid were not eligible under government ground rules. Another example: A. C. Rosander in his book, *Applications of Quality Control in the Service Industries,* reported that 25 percent of a bank's total operating costs were devoted to quality costs. He estimates that the PQC is broken down in the banking industry as follows (16):

Prevention cost	2%
Appraisal cost	28%
Internal error cost	41%
External error cost	29%

An even higher estimate of bank quality cost comes from William J. Latzko, then Quality Control Manager for Irving Trust Company, who reported in his paper "Reducing Clerical Quality Costs" that his studies of a sample department indicate that nearly 40 percent of the department's total operating cost was devoted to PQC. He went on to report that the PQC were distributed as follows (17):

Prevention cost	2%
Appraisal cost	28%
Internal failure (error) cost	41%
External failure (error) cost	29%

Frank Scanlon, Director of Quality, Hartford Insurance Group, reports, "17 to 25 percent of all the paperwork is being reprocessed due to errors. Errors in Accounts Receivable can impact profits; e.g., undercharging errors are seldom brought to the company's attention whereas overcharges usually are" (18).

As these numbers indicate, service industry poor-quality costs offer substantial opportunities for improvement.

7

Indirect Poor-Quality Cost

Poor Quality Costs Your Customers

Far too often we base our business decisions on the immediate impact they have on the business, ignoring the impact these decisions have on our customers and the long-range impact they may have on the business itself. Today if we want our company to have a future, we need to do more than just meet customer requirements; we need to fulfill customer expectations. Your needs for a hotel are fully met by a Motel 6 because they supply a clean room, a flat bed, and hot and cold running water. Why is it then that many traveling executives select a hotel like the Hyatt Regency? It's because they want more than just their basic requirements fulfilled; they want their expectations to be met. When you merely meet requirements your customers have no reason to return, but as long as you fulfill their expectations, you'll have them for a long time.

When it comes to products, many manufacturers try to make up for poor products with long warranty periods, reasoning that if the product fails, the customer is provided with a new one. Too many companies reason that it is too expensive to remove one defective unit in a lot of one hundred. Why not let your customer serve as the company's final inspector? Data has proved that if the item is not expensive, frequently the customer will just throw it away and never return for a replacement unit. You can't dispute this type of data, but what the data does not reflect is what other brand that customer buys the next time he or she needs your product, or how many other sales you lost because of the bad publicity your company received as a result of shipping defective merchandise. The White House Office of Consumer Affairs reports, "96 percent of the unhappy customers never complain about discourtesies, but up to 81 percent will not buy again from the business that offended them. In addition, the average unhappy customer will tell his or her story to at least 9 other people, and 13 percent will tell more than 20 people" (19).

Armand V. Feigenbaum, president of General Systems Company, reported in the February 1986 issue of *Quality Progress,* "Today, when a customer is satisfied with quality—when he likes what he buys—he tells 8 people. When he is dissatisfied, he tells 22. That's the hard arithmetic of quality's effect on sales in the American marketplace. Cutting the cost of quality can give quality-leveraged companies a competitive advantage of five cents on the dollar—as much as ten cents in some companies." (20).

Customers used to *want* quality. Today they *expect* it, and get very irate if they don't receive it. In today's environment, most families depend on two paychecks to meet their financial obligations. In these cases, both individuals' time is valuable to them. Any disruption caused by errors that takes away from their precious leisure time provides a high level of frustration and disturbance. The same is true in business. A malfunction in one computer in Japan shut down the country's banking system during a Saturday's busiest period. The cost of the repair was nothing compared to the money that was lost to the bank and the inconvenience to the bank's customers.

What was the true cost to the U.S. space program, the astronauts, and the reputation of the U.S. when the Challenger spacecraft exploded shortly after takeoff in the early part of 1986? Certainly it was far more than the cost of the replacement booster rocket.

Yes, there is a huge cost related to poor quality's impact on the customer that is not reflected in the company's ledger, but is reflected in your customer's wallet. Sure, you don't have to consider it now, and you will get by. But many of your competitors realize that the customer is king. These are the companies that consider the impact of poor quality on their customers. These are the companies that will capture the major portion of the business in the future. To succeed in business today, you must pay attention to these indirect poor-quality costs.

The Second Major Division

The poor-quality cost system has two major divisions. The first part is direct poor-quality cost, made up of controllable PQC (prevention and appraisal), resultant PQC (internal and external error cost), and equipment PQC. The other major part of the poor-quality-cost system is indirect PQC, defined as those costs not directly measurable in the company ledger, but part of the product life cycle PQC. They consist of three major categories:

- Customer-incurred PQC
- Customer-dissatisfaction PQC
- Loss-of-reputation PQC

Why Use Indirect PQC?

The best way to explain why indirect PQC is an important concept is to look at an example. In 1963, when IBM first started to use PQC as a management tool, we almost made some very costly errors. At first glance, it looked as if we were overspending on appraisal functions, so we developed a formula to check each step of the manufacturing test process:

$$\frac{\text{Field cost/defect}}{\text{In-house cost/defect}} = \frac{(T_F \times R_F) + P_F}{(R_P \div D_P) + P_P} = X$$

where

T_F = average field repair time

R_F = average field labor and burden rate

P_F = average cost of replacement part in the field

R_P = average in-plant labor and burden rate

D_P = average number of defects detected and repaired per test hour at the operation under evaluation

P_P = average cost of replacement part in plant

Essentially what we were trying to do, using internal and external error cost data, was to determine which of our test operations should be eliminated. If the results from plugging in the data for any operation indicated that X was less than 1, the test operation would be terminated as not being economically justifiable. One restriction on terminating the operation was that none of the defects found during testing could be classified as critical (safety defects).

The next step was to write a computer program, and input the program and variable data into the computer. The results of our first run indicated that 65 percent of the test operations could not be economically justified. Deciding that these results were highly impractical, we re-evaluated the computer program and input data, and discovered two interesting facts:

1. Overhead costs were much greater in plant than they were in field repair centers.

2. Since field centers repair only defective items, no time was charged to testing good products. Although the plant repair time was much shorter, it did not offset the time required to determine if the product was defective.

With these two factors heavily weighting the plant defect cost, it would have been very easy to make a serious mistake and terminate testing activities if the indirect poor-quality costs had not been considered.

Back to the drawing board we went, to estimate what it cost our customers each time a defect occurred. We then rewrote the program so that this cost was added to the "field cost per defect." When the same basic data was rerun, we found that all the operations were cost-justifiable. The next step was to perform an analysis of failure versus time of failure in the operations whose X factors were close to 1. When this data was entered in the computer, we found that we could reduce the time spent on five operations, saving 48 minutes of test time for each machine.

The preceding example vividly demonstrates why dealing with only the direct

PQC can provide you with only part of the total picture—and a misleading part. Remember the blind man who felt only the trunk of the elephant and from that day on described an elephant as a snake-like animal.

To make the proper decision, the total PQC curve must be considered. This means that the indirect PQC must be added to the direct PQC before the decision is made.

In today's environment, customers consider both the acquisition and the maintenance costs before they buy. They usually receive relatively little information on maintenance costs, so they base their decision to buy on their own experience or the experience of close friends. This is an unfortunate situation, because the reputation of a statistically sound product can suffer based on a single instance. Let me illustrate with a personal example: I once decided to drive from California to New York, and then on to Florida, and back to California, stopping for a conference in Cleveland. To help keep my five-year-old son quiet and content, I bought a TV designed to operate in the car. To ensure that the set wouldn't get damaged, I placed it on its styrofoam shipping base and secured it to the car. The first day went well. The picture was a little snowy and rolled some, but my boy was happy and so was I. The second day, as we left Salt Lake City, the TV went black. That wasn't too bad, for I thought the set had a good warranty and certainly we could do without it for the first third of our trip.

When I reached Cleveland, I checked the phone book and there was no authorized dealer for my imported TV set. My next step was to check some of the department stores downtown. The second one that I visited carried my brand. The store supplied me with the name of the authorized repair shop and general directions on how to find my way across town to the shop. After two hours, I found the repair shop, only to be informed by the repairman that he would require my warranty paper before he could start work. So back into the car went the unrepaired TV. (If you've ever set anything on styrofoam, you know that a very slight movement causes squeaks, and that TV squeaked all the way back to California. The more the TV squeaked, the more unhappy I became.)

Upon returning to California, I went to a repair shop, this time with warranty in hand. The service man accepted the TV and told me to return in two days. Three days later I returned and was informed that the part had to be ordered and would not be in for three more days. I waited five days and this time I called to be sure the set was ready and, of course, it was not—but the part was expected the next day. This went on for seven more days before I was able to pick up the repaired TV. (After this experience, I bought a tape recorder. The company that made the TV made a tape recorder with the best performance/price combination, but I bought another brand even though it cost more and didn't sound quite as good.)

The difficulty with the TV was merely a transistor failure, but the failure was expensive, particularly with regard to indirect PQC. This is illustrated by the following comparison of direct and indirect PQC:

Direct PQC (external failure)

1.	Transistor	$2.38
2.	Repair cost	16.00
	Total external failure cost	$18.38

Indirect PQC

1.	Time required in Cleveland to locate, go to, and return from the repair shop: 3.0 hours at $10/hour	$ 30.00
2.	Thirty-two miles traveled in Cleveland at 20¢/mile	6.40
3.	Three trips: 45 minutes each way to the TV repair shop in California at $10/hour (4.5 hours total)	45.00
4.	Three 24-mile round trips to the TV repair shop in California at 20¢/mile	14.40
5.	Loss of tape recorder sale, 10 percent of $298	29.80
	Total indirect PQC	$125.60

Note that there was almost a 7 to 1 ratio between indirect cost and external error PQC. This does not include cost for loss of service for the two-month period; nor does it include any consideration for my inconvenience. Neither was consideration given to loss of sales that occurred when I told my friends about my problem. Not a normal case? Right. But $89.20 of the indirect PQC would still have been involved even if the entire episode had occurred in California.

Indirect PQC varies from product to product, making it difficult to recommend general cost-reduction procedures. The costs can be reduced by carefully and systematically analyzing each product type or family of products as your customers see them. In my case, the company should address three problems: the failing transistor, customer service, and the warranty.

The failing transistor—If the transistor had not failed, the problem would have been prevented. (Minimizing early-life transistor-failure rates is beyond the scope of this book and will not be discussed.)

Customer service—When the repairman could not meet his commitments, he should have notified me. His phone call would have reduced the indirect cost incurred in California by almost 30 percent. The average indirect cost of a repair could be reduced even further if the serviceman would give the customer the opportunity of waiting while he performs a quick assessment of the problem. This assessment might have eliminated the need for a second trip.

Warranty—Not having my warranty papers eliminated any chance of having the set fixed in Cleveland, and probably would have necessitated my making a second trip in California. A simpler warranty that could be honored by any dealer would be the answer. Perhaps it could be as simple as stamping the sales date on the product at the time of purchase. A specified warranty time from that date could cover repairs anywhere in the nation.

Customer-Incurred PQC

Customer-incurred poor-quality cost appears when an output fails to meet the customer's expectations. Typical customer-incurred PQC are:

- Loss of productivity while equipment is down

- Travel costs and time spent to return defective merchandise

- Overtime to make up production because equipment is down

- Repair costs after warranty period is over

- Backup equipment needed when regular equipment fails

Figure 7.1 shows the same product curves that were used in Figure 3.1 with the indirect customer-incurred PQC added. Notice how the best interim operating point has moved to the right and how customer-incurred cost decreases as the total number of

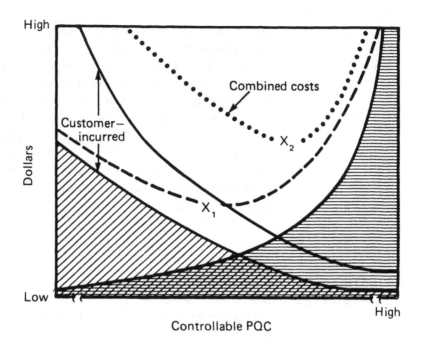

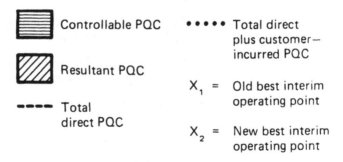

FIGURE 7.1 Addition of customer-incurred PQC to direct PQC.

errors decreases. It is not uncommon to see customer-incurred PQC exceed the total purchase price of a product during its life cycle. A good example is the repair cost during the life cycle of an old tube-type television, or one of today's video cameras. In Figure 7.2, the bar graphs that were devel-

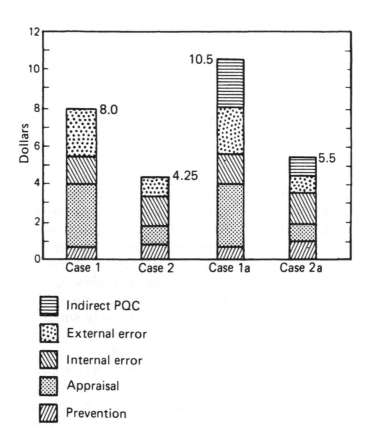

FIGURE 7.2 Best interim operating point with indirect PQC added.

oped in Figure 4.3 are shown. Directly beside them are two bar graphs, for Cases 1 and 2, where the customer-incurred PQC has been added to the original bar graph. In this example, a very conservative picture is portrayed; the customer-incurred PQC is only equal to the external error cost. Nevertheless, the figure vividly shows that reducing field error rates drastically improves total product PQC. This example also illustrates why it is imperative that you prevent errors from occurring—or, when you cannot eliminate errors, that you establish appraisal systems to detect them before the product is delivered to your customers.

Customer-Dissatisfaction PQC

Customer dissatisfaction is a binary thing. Customers are either satisfied or dissatisfied. Seldom will you find one who is in between. Figure 7.3 portrays customer-dissatisfaction PQC in terms of lost revenue versus product quality. On the left side of the curve, you will note a sharp decrease in lost revenue for only small improvements in product quality. This curve reflects the binary classification of satisfaction in the customer's mind. Once a customer's acceptance level has been reached, the curve becomes almost flat, even though the product-quality level continues to improve.

The quality level of U.S. and European products hasn't suddenly dropped. In fact, it has improved. What has happened is

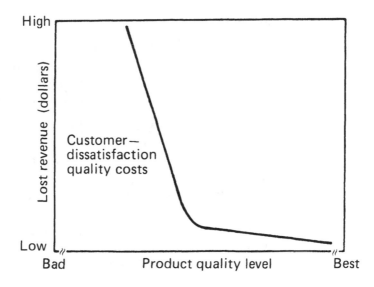

FIGURE 7.3 Customer-dissatisfaction costs.

that the customer-expectation level has changed. Customers now require a much better product to satisfy their expectations and demands. The customer-dissatisfaction level has moved to the right, but in many companies the quality level has remained constant, or has not kept pace with customer expectations (see Figure 7.4). These companies may very well have been making parts to specifications, but the specifications were not good enough to keep their old customers, let alone attract new ones. Many of our business leaders understand that customers' expectations are changing. For example, F. James McDonald, President of General Motors, has said, "At General Motors, we're proud

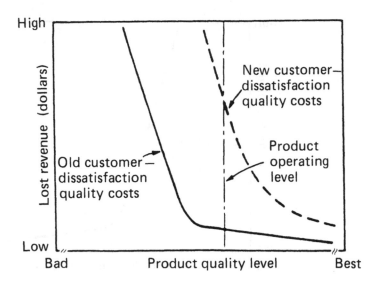

FIGURE 7.4 Changes in customer-dissatisfac-tion costs.

of the gains we've made in quality and in the productivity associated with the improvements, and our customers are verifying the results of our efforts. But even as our own performance improves, the expectations of our customers continue to rise. And there is no doubt that our competitors feel the same pressure" (7). Recognizing that our customers' expectations change and continuously resetting our standards is the only way we can keep our share of the market.

As the customer-expectation level increases without an accompanying increase in product quality, American and European countries begin to lose customers to Asian manufacturers. When you have satisfied customers, it is easy to keep them as long as your product remains price-competitive, or even if it is a little bit more expensive. The public tends to stay with the product that has a good record. But once you have lost your customers' confidence because of dissatisfaction with your product's performance, it is next to impossible to get them back into the fold. In addition, the "infection" tends to spread to family members and friends. A Gallup survey in September 1986 (26) showed that executives in leading U.S. corporations felt that poor quality's biggest financial impact was on lost sales (46%), putting it far ahead of rework (21%), scrap (14%), and repair (10%) costs. The use of customer-dissatisfaction PQC allows lost sales to be considered in the PQC equation. *When it comes right down to it, the customer's perception of our product is the most important fact.*

Figure 7.5 portrays what happens when customer-dissatisfaction PQC are added to direct and customer-incurred PQC. Again, the best interim operating point has moved to the right. This is a typical cost curve for a high-reliability product, such as a rapid-transit control computer or a car's braking system.

Figure 7.6 is a PQC curve for a product that does not require a high degree of excellence to satisfy the customer. This curve could represent a product like a change-making machine or a garden hose. The knee of the

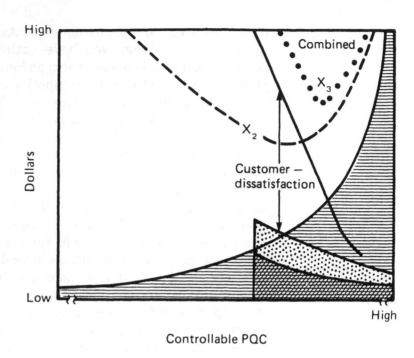

Controllable PQC

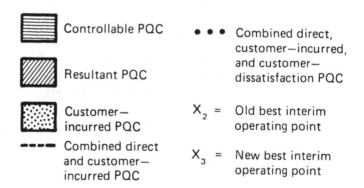

FIGURE 7.5 Customer-dissatisfaction PQC combined with direct and customer-incurred PQC.

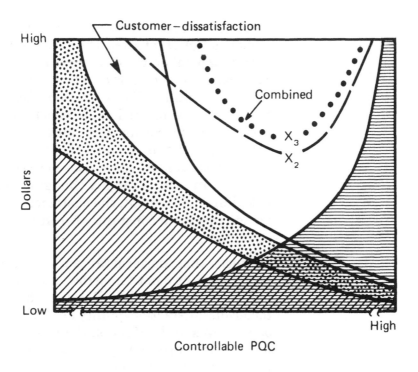

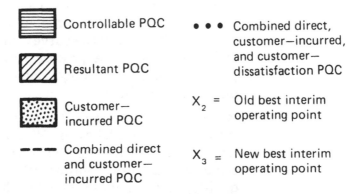

FIGURE 7.6 Cost curve for a product not requiring a high degree of excellence.

customer-dissatisfaction curve occurs at a higher defect level than the best interim operating point on the combined direct and customer-incurred curve. When this occurs, the customer-dissatisfaction PQC can be ignored because it has no effect on the best interim operating point. Just one caution in taking this action—the customer-dissatisfaction PQC curve does not remain constant. In today's rapidly changing technological environment, the customer-satisfaction threshold constantly becomes more stringent.

Products that perform at an acceptable level today may not tomorrow, and probably won't the next day. Because the customer-dissatisfaction quality level continuously moves to the right on the PQC curve, it must be continuously assessed, and action must be taken to decrease external error rates before the best interim operating point reaches the knee of the customer-dissatisfaction curve. The steep slope of this curve indicates that a product line that slips past the knee of the curve could be out of business before a process can be improved sufficiently to offset the customer-dissatisfaction factor.

Worldwide perception of U.S. product quality was very high following World War II. America had proved to the world that it could manufacture large quantities of high-quality, very reliable products. The world was our oyster and we were indulging ourselves in the bountiful rewards that were the result of the hard work we expended during the 1940s. In the post-war era we gained a mistaken feeling of manufacturing supremacy as the world toured our manufacturing plants to admire and learn from our manufacturing systems. Things looked bright and we began to rest on our laurels, but as a result productivity and quality growth rates fell off. By the beginning of the 1970s competition from Europe and Asia began to attract our customers away from our American-based companies. The quality level of American-made products continued to improve during the '70s, but in many cases it did not keep pace with our overseas competition and changing customer expectations. What happened was that many American-made products had slipped past the knee of the customer-dissatisfaction curve. In the 1980s, many American industries redevoted themselves to the need for quality output. This additional effort, as the 1985 ASQC Consumer Survey indicates, has paid off: The American consumer is regaining faith in American-made products (21). The magnitude of this change is clearly shown when a 1980 survey is compared to the one conducted five years later (22):

	1980	1985
Are foreign products equal to or better than U.S. products?	75%	33%
Will U.S. quality remain the same or get worse during the next 5 years?	69%	27%

In 1980, a survey (not exactly parallel to the one conducted in 1985) indicated that 75 percent of the people surveyed felt that foreign-made products were equal to or better than those made in the United States. In 1985, that had dropped to 33 percent. The earlier survey reported that in 1980, 69 percent of U.S. consumers felt that the quality of American-made products would remain the same or get worse over the next five years. In 1985, only 27 percent of Americans held those beliefs. These are very positive indications of a change in the correct direction.

Loss-of-Reputation PQC

Loss-of-reputation PQC are even more difficult to measure and predict than are customer-dissatisfaction and customer-incurred PQC. Costs incurred due to loss of reputation differ from customer-dissatisfaction costs in that they reflect the customer's attitude toward a company rather than toward an individual product line. The loss of a good reputation affects all product lines manufactured by a company. Its costs cannot be imposed upon an individual product PQC curve, but must be considered as a total effect on all product lines. It is for this very reason that it is considered good business practice to group and distribute products under different trade names based on expected performance.

Using Indirect PQC

How do you use indirect poor-quality cost? Some companies ignore it, others consider it only when they are making a change in the appraisal activities, and still others integrate it into their basic PQC system. The degree to which it is implemented is highly dependent on the importance that the company places on its customers. At the very least, some portion of the customer-incurred PQC should be added to the cost of each external error to show that your company realizes that the error has an impact on the customer. Management must decide on the percentage of the actual customer-incurred cost that is added.

There must be a continuous effort to evaluate customer expectations so that the output from the company remains on the right-hand side of the customer-dissatisfaction PQC curve. This analysis must be kept up to date because the knee of the curve continuously moves toward the right.

You cannot wait to react to customers' complaints, for even when a complaint is resolved you still can lose the customer. On an average, for every 10 complaints that are resolved, three customers will never buy from you again if they have an alternative.

8

Advantages of a Poor-Quality-Cost System

Why Use a PQC System?

A poor-quality-cost system by itself does nothing but upset management. It must be accompanied by an effective improvement process that uses the PQC data to direct its efforts and measure its progress. If used in this combination, a PQC system has a number of advantages, which will be discussed in the following paragraphs.

1. *PQC provides a manageable entity.* It changes quality from an abstract term into a manageable entity. When a nebulous, abstract term such as "poor quality" is placed alongside the harsh reality of price and schedule, it's very hard to compete for management's attention. Putting quality into a tangible unit (dollars) allows management to weigh the magnitude of the quality problems and give them correct priority. Poor-quality cost will provide management for the first time with a picture of

what the lack of quality is costing the company. The first reaction will be one of shock, then of disbelief, and eventually a realization that for the first time they have been provided a tool that allows them to properly manage an item that has major impact on the profitability of the company.

2. *PQC provides a single overview of quality.* It allows the total company quality situation to be summarized and viewed in one common term. Before a PQC system is implemented, quality reports talked about percent of lots rejected to supplier, yields, percent defectives at final test, defects per unit, mean time to failure, return during warranty, percent of control charts out of control, etc. Each area talked about quality using different terms that could not be summed together to provide management with a picture of the total quality problem. Often, the magnitude of the problem was masked because it was divided into many small parts and integrated across many different areas. When the many different reports, terminologies, and inputs are pulled together in terms of dollars, they can be accumulated, grouped, and summed together to provide a comprehensive picture of the company's cost of poor quality.

3. *PQC provides a means of measuring change.* It provides a way of finding the best interim operating point for the quality system. Management has many options on how to invest the limited resources that are available to them and they have many requests for funds. The poor-quality-cost sys-

tem provides a way to measure the return on quality investments and to adjust investments to meet today's changing needs.

4. *PQC provides a problem-prioritization system.* It separates the wheat from the chaff in helping the manufacturing engineering organizations prioritize their corrective-action activities. Frequently, the most visible problem or even the one that occurs most frequently is not the one whose solution has the biggest payback. The carpenter who is bending a hundred nails in building a house does have a problem, but that problem is not nearly as important to the contractor as a cement mixer left overnight with a full load in it. It is obvious from this example that frequency of occurrence is not the major factor in determining which problem solutions have the biggest payback.

It is also necessary to accumulate all the error costs related to a common problem from the time an item is received from a supplier until it is discarded by the customer. Frequently, common errors occur throughout the process that, when combined, represent a major problem.

5. *PQC aligns quality and company goals.* It assures that the quality goals are aligned with the company goals and objectives. Too often, enthusiastic quality professionals push for their own interest so enthusiastically that they forget the real reason the company is in business—to make a profit. Quality for goodness' sake alone can result

in providing a product that does not meet the customer's price requirements. Today, there really is a "big Q" to quality:

- Quality of performance
- Quality of price
- Quality of schedule

Product lines must have all three factors provided at a high level if they are to compete in today's marketplace. A quality product is one that functions when it is needed and is priced at a level that represents value to the customer. A quality product must support these requirements and, if it does, it will have a major impact on increasing the company's profits and increasing the company's share of the market.

6. *PQC provides a way to correctly distribute controllable poor-quality cost for maximum profits.* Today, most industries spend far too much of their controllable poor-quality cost on appraisal, but it should not be spent totally on prevention either. At some point you have to stop training people and put them to work producing usable output. The poor-quality-cost system provides a way to evaluate how effectively these critical dollars are being utilized.

7. *PQC brings quality into the boardroom.* It makes for effective communication between the quality staff and upper management. The complexity of quality terms like AQL, AOQL, SPQC, MTF, confidence limits, abnormal distributions, etc., are appropriate in the office of a statistician or a

quality engineer, but are as out of place in the boardroom as a person wearing a bathing suit. The top executives want to be involved in the quality system and know they need to set the example, but the quality professional needs to supply them with information that is understandable and can be used in any part of the business. Dollars is the common language of the blue-collar and white-collar workers. It is understood by the company president, the janitor, and the salesperson. Talk dollars to management; it makes good sense.

8. *PQC improves the effective use of resources.* Because poor-quality cost transforms quality problems into dollar-saving opportunities, the support groups (Quality Assurance, Manufacturing Engineering, Purchasing, and Product Engineering) have an effective tool to highlight problems that will provide the biggest payback to the company. Combining the poor-quality-cost data with a Pareto-diagram analysis results in a priority road map for the corrective-action process. In addition, it will tell you where, when, and how much to invest.

9. *PQC provides new emphasis on doing the job right every time.* When your employees stop looking at scrap as errors or defects and start looking at it as dollars, things start to change. There is relatively little impact on a person when he makes an error that causes a gear to be scrapped. But when that employee begins to think about it as a $50 bill that is being thrown away, it has a big impact on his future actions. The

department manager or supervisor is responsible for bringing about this change in thinking. This can be accomplished by reviewing the department poor-quality-cost report with the employees. In addition, a scrap display is another effective way to change the employees' thinking pattern. To start a scrap display, the manager simply takes scrap items and puts them out in a prominent location in the department, along with a sign that shows the cost of each item. It is very important to get the employees (both blue- and white-collar workers) to think about the financial impact their errors have on the company's bottom line.

10. *PQC helps to establish new product processes.* By having a financial understanding of the present process, future processes can be designed to eliminate high-cost appraisal and error operations. Frequently the poor-quality-cost data is used to justify new automated inspection equipment, reduce stocking areas, improve conveyor systems, or justify new process equipment. The only way you can eliminate errors is by focusing on the system, not the people. The people work within the limitations placed on them by the system.

11. *PQC provides a measure of improvement.* It provides the best measure of the effectiveness of the company's continuous improvement process. To compete today, a company must continue to improve its products and its services, both to itself and to its customers. To accomplish this, many

companies have implemented different forms of the improvement process. Quality circles, quality teams, quality of work life, participative management, zero defects, and statistical process control are just a few of the many methods that are being used to eliminate errors. Theoretically, all these programs sound good, and all *are* good if they are properly implemented and controlled. The question is: How do you know that they are performing the desired task? Well, when it gets right down to it, their sole purpose is to reduce poor-quality cost as a percentage of the total product cost. That is the only measure that can effectively evaluate corrective action and improvement processes.

12. *PQC reduction is one of the best ways to increase a company's profits.* Every poor-quality cost dollar that is spent takes away directly from the bottom line. Every poor-quality-cost dollar that is saved goes directly into the cash register.

Increased Profits

John Heldt, a consultant who has assisted many companies in implementing their poor-quality-cost systems, said, "Reducing the cost of poor quality will increase your overall profit more than doubling sales." He added, "Most companies are spending more than three times as much for poor quality as they are making in profit. Cut your poor-quality cost in half and you will at least double your profit" (23). How is that possible? Well, when you stop to look at the situation, there is a huge financial layout required to double production. More people, more equipment, more materials, more floor space, more support personnel, more sales personnel, more managers, more of just about everything. These expenditures detract from your profit. Sure, there are some advantages gained due to volume increases, but the large share of the sales price is used to fund the cost of producing the product. The profit margin changes only slightly. But for every dollar that poor-quality cost is reduced, a dollar is added directly to the profit margin. Table 8.1 provides a simple example.

In this example, if you were selling 10,000 units per month, the profit would be $140,000 when poor-quality cost is equal to 20 percent of the manufacturing costs. If you increased your sales to 20,000 units per week, your profit would be $280,000. By decreasing poor-quality cost from 20 percent to 10 percent, you would increase your profit per unit to $29 so the profit from 10,000 units would be $290,000, $10,000 more than the total profits realized from doubling the sales. In addition, you would have $3,700,000 tied up in producing 20,000 units at a 20 percent poor-quality cost and only $1,700,000 invested in the 10,000 units at a 10 percent poor-quality cost. This typical example vividly demonstrates why it may be far better to invest in establishing a poor-quality-cost system and installing an improvement process than it is to invest in a new plant, new equipment,

TABLE 8.1 Example of Costs and Profit at Two PQC
Percentages

	PQC at 20%	PQC at 10%
Cost of materials/components	$ 25	$ 25
Cost of labor including overhead	120	120
Poor-quality cost including overhead	30	15
Total production cost	175	160
Distribution and sales cost	10	10
Total cost to company	185	170
Sales price	199	199
Profit	14	29
Return on investment	7.6%	17%

and manpower to increase your output. Remember, often the customers for your increased output are simply not there. There is no doubt that reducing poor-quality cost is your best opportunity for increasing your profits.

Increased Market Share

There is another big advantage to improving quality. Today's sophisticated customers are very aware of the impact poor quality has on them, and they have adjusted their buying habits to their perceptions of the product's quality. No longer is the initial sales price the customers' major concern. Customers are willing to pay more initially to reduce their life-cycle cost. Reducing poor-quality cost has the advantage of providing better quality, which will increase the demand for your products, resulting in your company's capturing a bigger share of the market. The good thing about reducing poor-quality cost is that both the customers and the company gain.

Directions for Today and Tomorrow

PQC should be used as a tool to help management direct today's activities and plan for the future. It provides a measurement tool that helps them quantify how effective past activities were. It provides data that can be analyzed to pinpoint major problem areas. It provides the information needed to budget realistically. It also helps management prepare meaningful cost estimates for new products, services, or businesses.

What every company wants to do is maximize its return on investment. That means that management must have the ability to adjust its quality program to make maximum profit. As Figure 8.1 illustrates, if the quality of your product is low you will have extensive manufacturing and warranty costs and still not be able to obtain a premium price for your output. On the other side of the curve is a very high-quality output, but it uses up all the company's potential profit in obtaining its quality level.

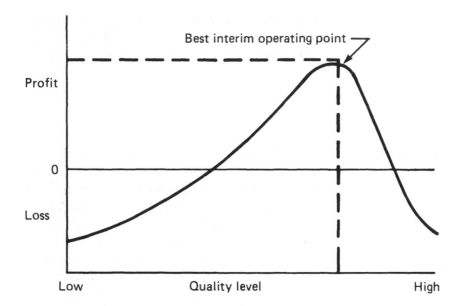

FIGURE 8.1 Optimizing profit through quality.

What management must do is adjust the quality system so that it operates at a point that provides maximum return on investment, keeping the long-term impact on quality in mind. A poor-quality-cost system is a tool that provides management with the information required to maintain this delicate balance.

In 1981, IBM's total poor-quality cost ranged from 15 to 40 percent of revenue. After IBM started the improvement process, John Akers announced, "We believe quality improvement will reduce quality costs by 50 percent over the coming years. For us that translates into billions of dollars and quality leadership in our industry" (4).

Summary

Poor-quality cost has become a valuable tool in directing the improvement activities of many corporations, large and small. The tool has been so widely accepted that it is now part of many of the government's major contracts. Companies such as IBM, General Electric, and American Telephone and Telegraph use it as a primary measure of the effectiveness of their efforts to improve quality and to integrate the quality responsibility into all functions of the corporation. It has also vividly demonstrated to management that they had been reacting to problems as they occurred rather than preventing them from occurring. As Ralph Wurster, editor of *Quality* magazine, wrote in June 1986: "I guess we're just too busy swatting flies to find out how they are getting in" (25).

Indirect poor-quality costs form the shadow over the poor-quality cost-analysis system being implemented in most companies today. Fortunately most systems operate to the left-hand side of the interim operating point. If your system operates to the right-hand side of the interim operating point on the combined direct poor-quality-cost curves, be sure that you fully consider the indirect costs before you cut controllable costs. A total poor-quality-cost system is as basic a quality tool as a micrometer. If you don't have a poor-quality-cost system, make this your goal for the coming year. If you have a poor-quality-cost system, use this effective tool of judgment. Trade in your opinion for facts.

> When you can measure what you are speaking about and express it in numbers, you know something about it, and when you cannot measure it, when you cannot express it in numbers, your knowledge is of a meager and unsatisfactory kind. It may be the beginning of knowledge but you have scarcely in your thoughts advanced to the stage of a science.
>
> —*Lord Kelvin*

Appendix: Typical Poor-Quality Costs

Prevention Costs

Prevention costs are all the costs expended to prevent errors from being made by all functions within the company, including development engineering, sales, product engineering, manufacturing, quality assurance, programming, personnel, etc.

Quality planning
  testing
  inspection
  audits
  process control

Education

Training procedures

Job-related training for employees

Equipment capability analysis

Design specification reviews

Product evaluation, characterization by product engineering

Customer interface to understand expectations

Technical manuals

Preproduction reviews

Defect-prevention activities

Early approval of product specifications

Computer-aided design (CAD)

Forecasting performance

Workmanship standards preparation

Preparation and documentation of training classes

Reliability PERT planning

Process definition

Postmortems

Modeling, simulation

Process characterization

Early entry (pre-analysis)

Manufacturing and quality engineering activities before design review

Design procedures preparation

Requirements reviews

Packaging qualification

Quality systems, procedures, and standards

Experimental activities (pilot production runs)

Reliability projection/prediction

Poor-quality-cost system planning

Installation reviews

Software planning

Software reliability projection/prediction

Systems analyst interrogating activities

Software documentation review

Preparation and review of system specifications

Flowchart review

Correlation analysis

Tape duplication and verification

Program quality plan

Test equipment plan

Automation planning to reduce defects

Work-flow layouts

Software engineering level controls

Process evaluation and characterization
by manufacturing engineering

Engineering product and process change
controls

Failure analysis

Activities to prevent an error from recurring

Customer surveys to detect changes in expectations

Forecasting and lead-time determination

Purchase cost targets

Schedule reviews

Budget generation

Expediting costs to assure proper deliveries (e.g., telephone bills)

Cost standards/estimates development

Process capability studies

Machine capability studies

Operator certification

Failure effects/mode analysis history data applied to new programs

Preventive maintenance

Process reviews

First-piece approval

Establishing quality reporting

Environment monitoring and control

Operator/inspector qualification

Preventive maintenance on noninspection and test equipment

Quality specifications for materials and processes

Routing sheets to control work flow

Review of supplier quality practices

Statement of requirements to suppliers

Supplier quotation system

Assistance to suppliers on quality training

Supplier qualification

Supplier quality evaluation survey

Procurement quality plans

Supplier quality training

Multiple sourcing

Appraisal planning

Planning personnel

Quality awareness activities

Long-range planning

Poor-quality-cost system development

Job descriptions

Instruction procedures/documents preparation

Operating manuals preparation

Establishing data acquisition and analysis

Quality motivation program

Security controls

Interviews with prospective employees

Quality improvement/excellence process

Zero-defect program

Housekeeping

Quality planning by all functions

Learning/understanding customer requirements

Surveillance of work areas

Department meetings

Environmental impact monitoring

Automation to improve quality

Scheduling work activities

Equipment correlation

Calibration of nonmeasurement equipment

Appraisal Costs

Appraisal costs are all costs related to the evaluation of already completed output and auditing to measure all functions' conformance to established criteria and procedures. It is all the costs expended to determine if an activity was done right every time.

Test and inspection materials (e.g., X-ray film)
Set-up for inspection and test

Product quality audits

Outside endorsements or approvals

In-process assessment

Process controls

Appraisal support

Test equipment records

Quality department administration

Training of quality assurance personnel

Training of inspection and test personnel in any area

System test costs

Invoicing review

Calibration/maintenance of production equipment used to evaluate quality

Product audits

Quality systems audits

Customer satisfaction audits

Outside lab evaluation

Life testing

Burn-in and stress analysis

Producibility/quality analysis review

Fault-insertion test

Training for special testing

Field performance testing

Walk-through analysis

Verifying workmanship standards

Tester monitoring

Maintenance and calibration/accuracy reviews

Installation appraisal costs (field testing/inspection)

Product qualification tests

Prototype inspection and test

Product design review

Drawing checking

Engineering audits

Product and process qualification/
evaluation by QA

Design/test specification reviews/analysis

Ledger review of profit/loss and balance
sheet

Time-card review

Capital expenditure review

Operating expenditure review

Product cost standards review

Financial reports, generation and review

Purchase-order review

Checking labor claims

Accumulation of cost data

Order-entry review

Production-rate review

Financial audits by outside corporations

Time required to work with outside corporations

Time required to do self-audits

Computer time
   program test
   function test
   performance test
   code verification

Data processing report review

Information systems cost associated with supplier rating

Data processing inspection and test reports

Review of test and inspection data

Safety review (operator safety)

Instruction procedures/document review

Total expenditure reviews (not delegating authority)

More than one manager's signature on a document

Source inspection of supplier plants

Process surveillance at suppliers

Supplier recertification

Incoming inspection cost

Employee surveys

Personnel appraisals

Internal audits of operation systems

Miscellaneous reviews

Operation audits

Upper management meetings with employees

Management meetings with customers

Processing security clearance

Security checks

Safety checks

Employee inspection of completed work

Internal Error Costs

Internal error costs are all costs incurred by the company related to errors detected before the output is accepted by the company's customer, or costs incurred before a product or service is accepted by a customer because all activities were not done right every time.

Installation failure costs

Downgrading (substandard product) cost

Overtime because of problems

Scrap or rework

Sorting activities

Reinspection because of rejects

Material review board action

Troubleshooting cost

Reinspection, retest cost

Failure verification/analysis and reports

Corrective action

Failure reports

Analysis of scrap

Analysis of rework

Failure support

Quality circles

Improvement teams

Activities that will keep an error from reaching the next customer

Redesign and engineering change (REA) cost

Terminated products

Engineering travel and time spent on problems

Modifications to process

Temporary (soft) tooling

Abandoned programs

Engineering change scrap and rework

Cost-reduction activities

Billing-error cost

Bad-debt turnover

Payroll-error cost

Salvage and defective material reports

Premium shipping costs to make up for late products

Inventory out of control

Incorrect-accounting-entry cost

Purchase-order-rewrite cost

All expediting costs

Supplier-cancellation cost

Overdue accounts receivable

Improper payments to suppliers

Poor-quality-cost reviews

Financial report corrective action cost

Off-spec/waiver (nonconforming material permit)

Supplier scrap

Rework of supplier parts

Excess inventory because of undependable supplier deliveries/quality

Losses because of supplier delinquencies

Shipping costs on return to suppliers

Trips to suppliers to resolve problems

Scrap/rework because of supplier faults

Retraining, rewriting/updating instructions/documents/invoices

Out-of-control condition (line-down costs)

Document changes

Relocations/moves/rearrangements not planned

Missed schedule

Equipment downtime

Redoing work (retyping, correcting errors, etc.)

Accidents, injuries

Overdue costs

Waiting costs (e.g., because of a meeting's not starting on time)

Keeping track of system failures

Theft

Absenteeism

Personnel turnover cost

Lateness (failure to respond)

Missed targets

Missed schedule costs

In-process inventory over one week's needs

Incorrect time estimates/records

Redundant equipment in case of failure

Handling employee complaints/grievances

Incorrect labor level

Loss of billing discounts

Efforts to fix blame

Time spent investigating nonexistent problems

Time lost because reports are wrong

Profit lost because reports are not on time

Unused reports

Time spent to follow up when schedules are missed

Disclosure of company secrets

Line-down cost because of parts shortages

Working around parts shortages

Retyping

Proofreading by someone other than typist

Misfiled information

Filing cost of unused documents

Time required to find equipment that works (e.g., another copier)

Clerical material scrapped because of errors

Trying to meet bad estimates

Drafting errors

A design's not passing review

Productivity loss or increased salary cost when time estimates are too high

Orders lost because bids were received too late

Replacing stolen assets

Lost time because work area is not laid out correctly

Lost sales because telephones are not answered promptly

Replacing tool and tester because engineering changes made them obsolete

Labor-utilization index less than 1

Time spent by higher-level people doing low-level jobs (e.g., managers making copies of documents or engineers typing letters)

Unused floor space

Change orders due to errors

Losses because of billing errors

Time required to correct time cards

Program-debugging time

Computer rerun costs

Lost sales because forecasts were too low

Equipment-utilization index less than 1

Doing things that don't need to be done

Time required to interpret poor reports

Processing a suggestion more than once

Not following procedures

Material delays

Damaged goods and equipment

Processing insurance claims

Loss of travel discounts

Preparing and evaluating rework procedures

Dismissing unsatisfactory employees

Loss of savings when suggestion investigation takes more than one week

Using out-of-spec items

Utilities not needed (e.g., lights left on) or used for rework

Sales lost because of stock shortages

Publications scrapped because they are out of date

Stopping production because of poor-quality output

Stopping production when there is no real problem

Development engineering activity that does not result in a new product

OSHA fines and cost of corrective actions

Redirecting mail incorrectly addressed

Fires

Late invoices

Added mailing or shipping costs because item was not ready in time for regular method

Added cost of rush orders because parts are not in stock when needed

Revising plans that are missed

Profit lost when product is not shipped because facilities were late

Waiting time when equipment is down

Lost assets

Payroll errors

Legal costs of defending the company

Rejected proposals

Errors in market forecasts (lost sales, overstock, unused facilities)

Lost savings because suggestions were not implemented on schedule

Personnel injuries

External Error Costs   External error costs are all costs incurred by the producer because the external customer is supplied with unacceptable products or services.

Canceling suppliers

Verifying failure

Field repair center, total expenses

Training repair personnel

Salary for repair personnel

Loss of rentals

Downtime charges

Product recall

Modification delays and costs

Handling complaints

Shortages of components or materials

Product or customer service because of errors

Products rejected and returned

Returned-material repair

Warranty expenses

Reinspecting and retesting

Troubleshooting

Corrective actions

Failure support by plant

Engineering change scrap and rework

Analysis of returns

Analysis of warranty

Direct customer contact on field problems

Redesign

Engineering-change analysis

Engineering time and travel on problems in field

Going back to customer to re-evaluate

Customer change requirements

Documentation changes

Field inventory

Handling returned material

Accounting cost related to returned items

Evaluation of field stock and spare parts

Overdue costs

Bad debts

Product liability suits

Theft

Loss of customer good will

Loss of sales because of poor service

Accidents/injuries

Costs due to waiting

Overhead for repair personnel

Malpractice suits

Redundant equipment

Lost sales due to poor output

Programming service to make changes

Field reports

Sales and service reports

Return and allowance reports

Returned-materials reports

Failure reports

References and Suggested
Additional Reading

References

1. James E. Olson, "The State of Quality in the U.S. Today," *Quality Progress,* July 1985, page 33.

2. Ronald Reagan, "A Presidential Proclamation," *National Consumers Week,* 1986.

3. Quoted in "World-Class Quality," *Quality,* January 1986, page 14.

4. John F. Akers, remarks at American Electronics Association Seminar on Quality, Boston, Massachusetts, March 13, 1984.

5. Richard K. Dobbins, *Quality Cost Management for Profit,* American Society for Quality Control Annual Quality Congress, 1975 Transactions, Milwaukee, Wisconsin.

6. O. G. Kolacek, *Quality Costs—A Place on the Shop Floor,* American Society for Quality Control Annual Quality Congress, 1976 Transactions, Milwaukee, Wisconsin.

7. Quoted in H. James Harrington, *Quest for Quality,* American Society for Quality Control, Milwaukee, Wisconsin, 1987.

8. Ronald Reagan, "A Presidential Proclamation," *National Quality Month,* October 1984.

9. John A. Young, "The Renaissance of American Quality," *Fortune,* October 14, 1985.

10. Personal communication, January 1986.

11. Philip B. Crosby, *Cutting the Cost of Quality,* Industrial Education Institute, Farnsworth Publishing, Boston, Massachusetts, 1967.

12. H. James Harrington, *Excellence: The IBM Way,* IBM Technical Report TR 02.1278, May 1986.

13. John J. Heldt, *Controlling Quality Costs: A Self-Study Guide,* MGI Management Institute, Harrison, New York, 1986.

14. J. R. Forys, "Redefining Quality Awareness," *Quality Progress,* January 1986, page 14.

15. Douglas D. Danforth, "A Common Commitment to Total Quality," *Quality Progress,* April 1986, page 17.

16. A. C. Rosander, *Applications of Quality Control in the Service Industries,* Marcel Dekker, New York, 1985.

17. William J. Latzko, *Reducing Clerical Quality Costs,* American Society for Quality Control Annual Quality Congress, 1980 Transactions, Milwaukee, Wisconsin.

18. Frank Scanlon, *Cost Reduction through Quality Management,* American Society

for Quality Control Annual Quality Congress, 1980 Transactions, Milwaukee, Wisconsin.

19. White House Office of Consumer Affairs, in a bulletin on consumers' attitudes, 1982.

20. Armand V. Feigenbaum, "Quality: The Strategic Business Imperative," *Quality Progress,* February 1986, page 26.

21. "ASQC/Gallup Consumers Survey, *Quality Progress,* November 1985, page 12.

22. H. James Harrington, "Planning the Future," *Quality Progress,* February 1986, page 53.

23. John J. Heldt, personal communication, October 14, 1986.

24. John F. Akers, remarks to security analysts, Boca Raton, Florida, as reported in *Wall Street Journal,* March 16, 1984.

25. Ralph Wurster, "New Idea?? Not Now!," editorial comments, *Quality,* June 1985, page 5.

26. Gallup Organization, Inc., "Executives' Perceptions Concerning the Quality of American Products and Services," September 1986.

Suggested Additional Reading

Aerospace Industries Association, Washington, D.C., Quality Resources Study Annual Report (available only to members).

Armand V. Feigenbaum, *Total Quality Control: Engineering and Management,* McGraw-Hill, New York, 1961.

A. F. Grimm, *Quality Costs: Ideas and Applications,* American Society for Quality Control, Milwaukee, Wisconsin, 1983.

H. James Harrington, *The Improvement Process—How America's Leading Companies Improve Quality and Productivity,* McGraw-Hill/American Society for Quality Control, New York and Milwaukee, Wisconsin, 1987.

R. E. Heiland and W. J. Richardson, *Work Sampling,* McGraw-Hill, New York, 1957.

J. M. Juran, *Quality Control Handbook,* revised third edition, McGraw-Hill, New York, 1974.

Quality Cost Technical Committee, American Society for Quality Control, *Guide for Reducing Quality Costs,* Milwaukee, Wisconsin, 1977.

Quality Cost Technical Committee, American Society for Quality Control, *Principles of Quality Costs,* Milwaukee, Wisconsin, 1986.

Index

Milton Keynes UK
Ingram Content Group UK Ltd.
UKHW040100071024
449327UK00019B/698

9 780367 451516